Over Our Heads

Heads

A Local Look at Global Climate

John C. Ryan

With research assistance by
Chris Cramer
Angela Haws
Joanna Lemly

NEW Report No. 6
November 1997

Northwest Environment Watch
Seattle, Washington

In memory of Cary Lu

NEW thanks editors Ellen W. Chu and Sherri Schultz and reviewers Richard Gammon, K. C. Golden, James Kerstetter, Molly O'Meara, Rhys Roth, Sam Sadler, and Amy Snover. For their dedicated assistance, we also thank interns Yoram Bauman, Carolyn Beaver, Christopher Crowther, David Kershner, and Meredith Wenger and volunteers Peter Carlin, Norman Kunkel, Kirk Larsen, Lyn McCollum, Michelle Olsufka, John Roullier, Marilyn Roy, Erik Schweighofer, Sandra Singler, Amy Snover, Scott Stevens, and Darrel Weiss. NEW is grateful to its board of directors: Lester R. Brown of Washington, D.C.; Sandi Chamberlain of Victoria, B.C.; Aaron Contorer of Kirkland, Wash.; Alan Thein Durning, Sandra Blair Hernshaw, and Tyree Scott of Seattle; and Rosita Worl of Juneau.

Financial support for this book was provided by contributors to Northwest Environment Watch. These include more than 1,500 individuals and the Bullitt Foundation, C. S. Fund, Nathan Cummings Foundation, Ford Foundation, William and Flora Hewlett Foundation, Merck Family Fund, True North Foundation, Turner Foundation, and an anonymous Canadian foundation. Views expressed are the author's and do not necessarily represent those of Northwest Environment Watch or its directors, officers, staff, or funders. Northwest Environment Watch is a 501(c)(3) tax-exempt organization.

This book was printed in Vancouver, B.C., using vegetable-based ink and recycled paper bleached with hydrogen peroxide. Text: 100 percent postconsumer waste; cover: 50 percent preconsumer waste. Design by Visible Images, Seattle.

Excerpts from this book may be printed in periodicals with written permission from Northwest Environment Watch, 1402 Third Avenue, Suite 1127, Seattle, WA 98101-2118; tel.: (206) 447-1880; fax: (206) 447-2270; e-mail: new@northwestwatch.org; Web: http://www.northwestwatch.org.

Cover photographs: Storm cloud in Idaho © 1997 by Gary Braasch; B.C.'s Tatshenshini River by John Ryan.

Copyright © 1997 by Northwest Environment Watch
Library of Congress Catalog Card Number 97-76494
ISBN 1-886093-06-7

TABLE OF CONTENTS

It's in the Air 4

What's Coming Down *Impacts* 9

What Goes Up *Emissions* 32

What We Can Do · *15 Steps to Climate Leadership* 54

Notes 68

It's in the Air

TAKE A DEEP BREATH as you read these words. The air you inhale is mostly nitrogen, but it also contains smaller amounts of oxygen, water vapor, and argon as well as traces of other gases and particles. If you're as lucky as I am today, it will have a hint of salt spray blowing in from the sea.

The air around you also contains about a third of a percent of carbon dioxide: CO_2. While you inhale to draw oxygen into your bloodstream, your existence depends no less on the carbon dioxide you breathe. Such a minute concentration of an invisible gas not only sustains the world's plant life but also helps keep the Earth's surface at a livable temperature. Without the heat trapped by "greenhouse gases" like carbon dioxide, the Earth would be an icy, lifeless ball, 60° F (33° C) colder than it is now.

Though you'd never notice the difference, your breaths differ from those of your ancestors: over the past century and a half we humans have fundamentally changed our planet's air. When they were born in Germany, Ireland, Switzerland, and various United States, my grandparents' grandparents breathed air that had about 280 parts per million of carbon dioxide. Yours did too, wherever they lived. By contrast, the air blowing through my office window five blocks from the Seattle waterfront has about 360 parts per million. The air you're breathing does too, and that's more CO_2 than at any time since the origin of the human species. With the billions of tons of CO_2 and other greenhouse gases pumped by our industries, cars, and farms into the sky, we've increased the heat trapped by the atmosphere. We've raised the Earth's average surface temperature at least half a degree Fahrenheit (0.3° C) and apparently begun to change our climate in other ways as well.[1]

The ten warmest years since record keeping began in 1850 have all occurred since 1980. Global sea level has risen 4 to 10 inches (10 to 25 centimeters) over the past century. Spring now arrives one week earlier in the Northern Hemisphere. Heavy downpours have become more common in the United States. Some changes have been dramatic—like glaciers and ice sheets shrinking in both hemispheres—but to date, most have been subtle, barely noticeable above the year-to-year variations that occur naturally.[2]

Right now you might want to take another deep breath. It might help you prepare for the shock of what's coming. Humanity has only begun to feel disruptions to our climate; we can expect them to reach dangerous proportions if the world continues to rewrite the chemistry of the skies. Costly summer droughts and winter floods would befall the Pacific Northwest. Forests and rangelands would be seriously disrupted by pest and fire outbreaks; natural areas would unravel. Elsewhere, millions would become refugees as tropical lowlands and entire island nations disappeared under a rising Pacific Ocean. These calamities might not hit with full force until late in the twenty-first century; they might hit much sooner. Scientists expect the climate here in the Northwest to warm another 2° F (1° C)—approaching levels not seen in 10,000 years—within the next 20 years.[3]

That said, my aim is not to overwhelm you with a sense of dread. The sky is not falling. It's just heating up. A little. The trouble is that on a planet of such exquisite interdependence—where a degree's difference in average air temperature can be enough to cause grasslands to grow instead of forests and a degree's difference in water temperature can tip the scales against salmon racing their biological clocks to the sea—a little heating goes a long way.

My aim in this book is to combat complacency and help others defend our climate. How to stabilize our climate is straightforward enough: stop adding greenhouse gases to it. Yet doing so would require a fundamental transformation in our fossil fuel–based civilization. The good news is that political, not technical, obstacles prevent us

Does My Breathing Hurt the Planet?

It's *not* a ridiculous question. But the answer (thankfully) is no. Your body combines oxygen from the air and carbohydrates from food to generate energy, carbon dioxide, and water. When you exhale, you add carbon dioxide to the atmosphere. But the carbohydrates that fuel your respiration came from plants (or animals that ate plants) that removed carbon dioxide from the air. So by breathing, you are simply recycling carbon dioxide that was removed from the atmosphere by food you ate. When it comes to the climate, you don't have bad breath.

from using the atmosphere responsibly. Though it may seem farfetched, weaning ourselves from fossil fuels is an entirely feasible transformation—indeed, it is inevitable in the long run. And, most likely, reducing our greenhouse gas emissions will save us money compared with doing nothing. It will also make our cities and nations more efficient places to make a living and more pleasant places to live. The main problem is getting enough people, governments, and businesses to wake up and smell the global warming.

This book explores a global issue—climate change—from a local perspective. The locale is the Pacific Northwest, defined here as the watersheds of rivers that run through North America's temperate rain forest zone. Stretching from Alaska's Prince William Sound to the California redwoods and from the Pacific Ocean inland as far east as the continental divide, the Pacific Northwest encompasses most of British Columbia, Idaho, Oregon, and Washington as well as western Montana, northwestern California, and southeastern Alaska (see map inside front cover). This region covers 1 percent of the Earth's land surface, supports a $300 billion economy, and is home to 15 million people. With 0.25 percent of the world's population, we northwesterners produce more than three times our share of the world's greenhouse gases.[4]

Clearly, no nation or region acting alone can protect the global climate. But the world is waiting for its worst polluters—namely, the United States and Canada—to lead it away from the climate catastrophe they are leading it toward. In a world hungry for the know-how to use energy efficiently and to tap into the immense energy flow available from the sun, the Northwest is poised to lead—

and to prosper by leading. With its collection of companies on the cutting edge of climate-friendly technology, its history of forward-thinking legislation on energy conservation, and its low reliance on fossil fuels compared with other regions of the continent, the Pacific Northwest can help transform North America from an unmatched polluter of the sky to a pioneer in defending it.

The Northwest continues to generate greenhouse gases in record amounts, yet a better way can already be glimpsed in the region. In 1997, for example, Oregon's legislature adopted the United States' first law controlling carbon dioxide emissions. Cities across the region have begun devising strategies to reduce their atmospheric impacts. And individuals and families are simplifying their lives and, often unwittingly, helping protect the climate.

Northwesterners actively defending the atmosphere remain small islands in a sea of sport utility vehicles and ill-advised public policies. Few North Americans are concerned about climate change, even though nothing is more important to the world's well-being than the stability of its climate. And nothing is more important to climate stability than North America's ability to slash its greenhouse gas emissions.

Backing away from the uncontrolled experiment we humans are conducting on the Earth will take unprecedented effort. But if you doubt the effort will be worthwhile, look around you. Climate made the Northwest what it is: a place of tall trees, salmon streams, and human communities that have prospered beyond the wildest dreams of our grandparents' grandparents. To protect this place, we need to protect the air over our heads.

What's Coming Down

Impacts

AT THE SNOWBANK'S EDGE, willows and bright purple penstemon flowers lean out over a trickle of ice-cold water. It's a hot summer day on the south flank of Mount Hood, and it hasn't snowed for months. Yet so much snow falls here each winter that some remains on the slopes, melting and retreating upslope day by day, until the end of summer. Just uphill begin the permanent snowfields and glaciers ringing Oregon's highest peak.

I've come here, to snowline in the Cascades, because reading climatological studies is one thing, but seeing for myself where global warming will hit is another. Standing on top of the gritty snow, I can hear running water gurgle below my feet. Where the gurgle hits daylight, I can see water dripping into the unnamed rivulet that tumbles down the mostly barren mountainside high above treeline. The rivulet will join others like it to form snowmelt creeks and rivers and eventually the Columbia, the largest river on the west coast of the Americas. Most of the water in the Columbia as it passes through the Oregon and Washington Cascades started out the same way: as drips of melting snow.[5]

Before the trickle at my feet reaches the Columbia, it will merge with others at the headwaters of the White

River and dive through the boulders and glacial dust of the broad White River Canyon separating the Timberline and Mount Hood Meadows ski areas. Below the canyon, the water will keep streamside pines and salamanders cool and wet in the dry forests of the eastern Cascades. It will rush down into the semiarid valley of the Deschutes River, where it will water spawning beds for steelhead trout and perhaps irrigate cherry orchards near the confluence with the Columbia. Once it joins the Big River, the water will generate electricity at The Dalles and Bonneville Dams and help carry young salmon to the Pacific.

The snow I see on Mount Hood, like that on other Northwest peaks, is worth a lot of money—and not just because skiers pay to play at the peak's resorts or because it adds a scenic quality to life in the Portland area. You might not believe it, but a major winter storm that dumps snow in the Cascades adds millions of dollars to the regional economy. That snow will melt slowly over the dry Northwest summer, supplying the region with river water—and all the benefits that water provides—when it would otherwise have little. If the air were slightly warmer when that same storm hit, it might snow on the highest peaks but deluge most of the Cascades with rain. The water would run off rapidly, and the storm could cost the region millions of dollars in damaging floods and lost hydropower. It might also take a toll of human lives.

Rapid climate change will disrupt the natural and human communities of the Pacific Northwest in many ways. Among other impacts, it will harm the forests and alpine meadows that spread for miles around me from high on Mount Hood. It will threaten rare species like the spotted owls and marbled murrelets that I know live inside these woods. But to see the most important effects of a changing climate, I need only look at the tiny stream at my feet. Water—as snow, rain, streams, rivers, reservoirs, and the sea—is central to life in the Pacific Northwest. Disruptions to its fall and flow have a central role in the unfolding story of what climate change will do to this place.

TO UNDERSTAND what will really happen here, you have to start with the experts, the climate modelers with their supercomputers. Atmospheric scientists have rapidly improved their ability to measure and model how the world's climate responds to greenhouse gases. By 1995, the Intergovernmental Panel on Climate Change (IPCC)—the preeminent scientific body representing 2,500 of the world's atmospheric scientists—could state authoritatively that humanity has caused at least part of the warming observed globally since the turn of the century. The panel concluded, in the cautious language of consensus science, that "the balance of evidence suggests a discernible human influence on global climate."[6]

The warming has been geographically uneven, generally more rapid near the Earth's poles and far from its oceans. Average temperatures over Idaho, Oregon, and Washington have risen 0.5° F (0.3° C) over the past century; in British Columbia, temperature increases over the past century range from 0.7° F (0.4°C) along the coast to 1.4° F (0.8° C) near the Yukon. These changes are small compared with the wider year-to-year swings in average temperatures. As climate change unfolds in the coming years, however, human-induced warming will become more pronounced compared with natural fluctuations.[7]

Scientists use complex computer models to simulate our planet's response to increased greenhouse gas concentrations. Given the even greater complexity of the real world, the models can only paint broad-brush scenarios of expected change. The IPCC projects that, if carbon dioxide levels in the atmosphere double (as they are expected to by the middle of the twenty-first century), average temperatures at the Earth's surface will likely warm 1° to 3.5° C (2° to 6° F), sea level will rise 6 to 36 inches (15 to 95

Will Climate Change Hurt My Breathing?

When fossil fuels burn, oxygen from the air bonds with carbon in the fuel to make carbon dioxide. For every pound of carbon added to the atmosphere, about 3 pounds of oxygen are consumed. Does this mean we're running out of oxygen? No. There's no shortage of oxygen over our heads—it's one-fifth of the atmosphere, about 60 times more prevalent than CO_2. Even if we burned enough carbon to triple the amount of CO_2 in the air, the world's supply of oxygen would barely be dented.

Of course, if we tripled the CO_2 in the atmosphere, we'd have more greenhouse gases than the planet has seen in more than 50 million years, and a broad array of ecological disruptions as well. Even if CO_2 is "only" doubled, warming is expected to worsen forest fires and heighten the formation of ozone smog in many urban areas. Both changes could make breathing more difficult—especially for asthma sufferers, children, the elderly, and others sensitive to air pollution.[8]

centimeters), and increased heat will speed the Earth's hydrological cycle of precipitation and evaporation. These changes may sound small, but temperatures during the last Ice Age, when Canada and the northern United States were covered in a mile-thick sheet of ice, were only 5° to 9° F (3° to 5° C) colder than today. A warming of even 1° C per century would be faster than any occurring in the last 10,000 years.[9] **Temperatures rapidly rise.** Global models are especially imprecise at the regional level. But they do suggest a troubled future for the ecosystems and communities of the Pacific Northwest. Climate scenarios used by scientists at the Joint Institute for the Study of the Atmosphere and Ocean (JISAO) at the University of Washington project a 2° F (1.1° C) warming by 2020, a 4.5° F (2.5° C) warming by 2050, and further warming after that for the U.S. Northwest. Warming is expected to be somewhat greater farther north. Environment Canada projects that doubled carbon dioxide concentrations will cause 5° to 11° F (3° to 6° C) warming in British Columbia after 2050, with greater warming in the interior than on the coast. For all the Northwest, scientists expect more precipitation and greater warming in the winter and less of each in the summer.[10]

Our climate is changing in unprecedented ways, taking us into uncharted waters. Surprises are in store as we leave the known realm of our historical climate. We will never know exactly what will happen next. Climate is unlikely to respond smoothly and predictably: in the geological record, climate has passed certain thresholds, then changed suddenly and unexpectedly. The impacts described here should be read not as predictions but as *plausible* outcomes if we do not greatly curtail our greenhouse gas emissions. Without such reductions, serious consequences are likely to befall the Northwest in the next 25 to 50 years. We may experience the effects of smaller-scale change even sooner.[11]

THE PACIFIC NORTHWEST has two basic types of rivers: coastal rivers, whose flows begin mostly as rain; and interior or high-elevation rivers, which are fed largely by melting ice and snow. Runoff from coastal rivers peaks with winter rains, whereas the runoff from interior rivers peaks in late spring and summer as warm air melts the snowpack. Under the altered climate projected for the Northwest—warmer, wetter winters and warmer, drier summers—both types of rivers will become more prone to summer droughts.

If winter temperatures in the Northwest warm 5° F (3° C), freezing level (the borderline between rain and snow) should move about 1,700 feet (500 meters) upslope. That's enough to put major ski areas like Washington's Snoqualmie Pass and Vancouver's north shore resorts out of business. Because of the basic triangular shape of most mountains—the higher you go, the smaller they get—rising snowlines will mean dramatically less total snowfall. A 500-meter rise would shrink the area of winter snowpack in the Columbia Basin nearly in half. (To picture this change, look at the snowcapped mountains on the horizon in winter; now picture them covered in only half as much white.) Snowfall will probably increase at higher elevations, but not enough to offset the dramatic reduction in snow-covered area. The overall effect would be a large decrease in spring and summer snowmelt.[12] **Snow turns to rain.** In addition to shifts in the timing of runoff, the annual average flow of the Columbia could be reduced up to 15 percent in the next 20 years. Reduced flows would be most pronounced during the low-flow season, August through January. Lower water levels would reduce hydropower production and put additional pressure on hard-hit salmon populations: the Columbia currently has sufficient flow to move endangered salmon safely past McNary Dam 85 percent of the time, but it would have that volume only 76 percent of the time by the year 2020 under the JISAO scenario. Just upstream on the Snake River, the flows past Lower Granite Dam would be sufficient for salmon less than half the time.[13]

Low-flowing rivers would also put the pinch on farmers—who withdraw far more water from the region's streams than all other users combined. Because higher temperatures boost plants' thirst, the need for irrigation regionwide will likely increase even as rivers have less water to give up. Conflicts over already scarce water supplies in the inland Northwest will heat up with the climate. **Lower rivers hurt farmers, fish.** Especially in heavily tapped streams like the Klamath, the Snake, and the Yakima, climate change could also worsen pollution: less water would remain to dilute fertilizers and soil running off croplands—the region's largest source of water pollution.

Lower summer flows would cause higher water temperatures, which are especially harmful to cold-water fish like salmon. In the Fraser—B.C.'s largest river and the world's most important salmon stream—unusually high river temperatures in 1994 apparently killed half a million sockeye salmon before they could reach their spawning grounds. Warm ocean conditions also harm salmon, in part by allowing warm-water predators like mackerel to expand their range. The governments of B.C. and neighboring U.S. states are now feuding over whose fleets are overfishing their shared salmon stocks, but none is rushing to fight what may be a greater enemy of cold-water fish: warm water.[14]

A THOUSAND MILES northeast of Mount Hood in the Canadian Rockies, glacier lilies and spring beauties carpet the ground next to mountaintop snow and ice in the upper reaches of the Columbia River basin. Before the river passes below Mount Hood, snow and rain landing on British Columbia and seven U.S. states will feed it. But the frozen stuff atop the Canadian Rockies is some of the most valuable anywhere: in addition to all the benefits it provides naturally, it will generate electricity at 14 dams on its way to the Pacific.

Only one-sixth of the Columbia River basin lies in Canada, but nearly a third of the river's annual flow begins there—more than from Idaho and Montana combined. The Canadian flow of the Columbia depends not just on snow but on ice: the hundreds of glaciers in the Columbia and Rocky Mountains. In late summer, up to 90 percent of the river's flow just before it enters the United States is runoff from glaciers. These water sources are especially threatened by climate change: glaciers from the Oregon Cascades to the Canadian Rockies are shrinking, and losses are expected to accelerate in coming decades. In Alaska

and northern B.C., snowfall is expected to increase enough to more than offset the effects of higher temperatures. But south of there, glaciers are projected to continue to retreat, some quite rapidly.[15] **Glaciers dwindle.**
Montana's Glacier National Park may soon need a new name: "the park formerly known as Glacier." Since 1850, 100 of its 150 glaciers have disappeared, with the warming trend accelerating in the past 30 years (see photos). At present rates of warming, its remaining 50 ice fields will disappear completely within 30 years. Dan Fagre, a U.S. Geological Survey researcher stationed at the park, puts it succinctly: "Basically, we have pathetic remnants of glaciers here, and they're dying fast."[16]
Initially, increased runoff from melting glaciers could offset the effects of reduced snowfall on stream flows. But once a glacier is mostly gone, its runoff will decrease drastically, within only a few years, and stream flows will be similarly slashed. Hydropower production throughout the Columbia River basin, and especially at dams in southeastern B.C. and elsewhere near the basin's glacial headwaters, could be dramatically curtailed.[17] **Energy production drops.**
When retreating glaciers break apart, they can cause sudden floods and rock avalanches. Few people will be directly affected by these events, since almost no one lives beside an active glacier. Like most Northwest residents, I live in the lowlands. But our lowland homes are not safe from floods and landslides. Coastal rivers' floods are likely to become more severe and frequent as winter precipitation increases and the share arriving as rain instead of snow rises sharply.[18]

Ice Cave, Boulder Glacier, Glacier National Park, Montana, July 1932

Same location, 1988

Powerful winter storms hit the Northwest several times in the 1990s, causing severe floods and thousands of landslides from California to British Columbia. Though these events cannot be ascribed to global warming, they give an indication of what is in store as our climate changes. **Winter floods worsen.** The 1996–97 floods were especially damaging: tens of thousands of northwesterners were evacuated, and several in rural Oregon and the Seattle area lost their lives as landslides triggered by heavy rains demolished their homes. Total damages in the Northwest states were estimated at $3 billion. In Idaho, mudslides and flooding stranded more than 10,000 people, and road repair alone cost $10 million. The same winter storm caused an estimated Can$200 million (US$150 million) of damage in southwestern B.C.[19]

The Northwest has always had winter floods, and these would happen even if our climate were not changing (see sidebar). But with global warming, floods in much of the region will become more severe and numerous. Changes in the frequency and intensity of flooding would harm aquatic species that have evolved within a certain range of stream conditions. Extreme floods can, for example, wash away young salmon, salmon eggs, and even the gravel beds salmon use for spawning.[20]

The increase in extreme rainfalls likely to come with climate change may warrant more concern than changes in average conditions. **Heavy downpours increase.** In February 1996, a deluge of warm rain hit the snowpack in the Cascades, unloading a double whammy of rain and snowmelt into the region's streams and rivers. In places where the flow was great enough and the land steep or

El Niño: Vary Important

Climate will always vary naturally from year to year. The fluctuations can be dramatic, as happens when an El Niño–Southern Oscillation event ("El Niño" for short) brings unusually warm air and water to the Pacific Northwest. El Niños often wreak havoc with fisheries and other important natural resources. In 1997, one of the strongest El Niño years on record, two Washington fishermen pulled in a 7-foot marlin—a warm-water fish usually found off the coast of Mexico and the first ever caught off the Washington coast. Sunfish, blue sharks, and other strange subtropical fish were caught in the Gulf of Alaska. Surfers even played in Washington waves without wetsuits.

Climate will be no less variable with global warming: there will be cold years and warm years, dry spells and wet spells. It's simply not possible to point to a given storm, or even an unusually warm year, and say that it was caused by global warming. Even several years in a row of "abnormal" weather may be only a result of an extended El Niño cycle, not global change. But global change could increase the intensity and frequency of El Niños and other natural oscillations or, at the least, combine with natural variation, bringing record-breaking conditions, and even marlin, to the Northwest.[21]

By pumping out greenhouse gases, we are tilting the scales in favor of warmer temperatures, wetter winters, and drier summers and increasing the odds of extreme conditions. We will have fewer extremely cold days and years and more extremely warm ones. The averages will shift, even as we are left guessing just what each day's or year's weather will bring. As researchers like to say, climate is what you expect; weather is what you get.

unstable enough, the floods became "debris flows"—water mixed with mud, boulders, even trees rushing downhill at up to 30 miles per hour, ripping a broad swath out of streams, forests, and roads in its path. Such debris flows can produce more erosion and sedimentation overnight than decades of smaller storms do. Gordon Grant, a U.S. Forest Service hydrologist studying streams in Oregon's Willamette National Forest, explains hydrology in the Cascades as "decades of boredom punctuated by hours of chaos." Commenting on the aftermath of the February 1996 storm on a stream scientists have been monitoring for decades, he noted, "We've learned that more sediment comes down during a couple of hours during a flood than for the remaining 40 years combined."[22]

The same storms would have caused many fewer landslides had the Northwest's watersheds not already been hammered by decades of road building and clearcutting. But that's the point: climate change will *not* descend upon a pristine landscape; it will be an added stress on ecosystems already seriously compromised by human activities. Natural systems are already losing much of their diversity because of habitat destruction, pollution, and other human impacts. The combination of factors will do more damage than all of them acting separately could accomplish.

FROM MY SNOWY PERCH high on Mount Hood, I can see all the way to the snowcapped cone of Mount Jefferson 50 miles away and beyond. Except for a few clearcuts, a highway, and a ski resort or two, the landscape below me is almost entirely forested. It is not an accident that the landscapes of the Northwest are dominated by evergreen trees: it is a result of our climate. Wet winters and dry summers give conifers, which grow year-round, an advantage over deciduous trees. If our climate changes rapidly, forests, grasslands, and other Northwest ecosystems will be profoundly altered.

As temperatures rise, species will be forced to migrate uphill or northward to maintain their preferred conditions. With a 5° F (3° C) warming, summer temperatures normally found at the Oregon-Washington border would be found near Kelowna, B.C., about 280 miles (450 kilometers) north. Seattle could have a climate roughly like that of Eugene, Oregon. As a Seattleite, I might think Eugene's warmer climate would be highly agreeable, but local wildlife and ecosystems would probably disagree. Species have differing abilities to disperse (the winglike seeds of maples are carried by the wind, for example, while conifer seeds grow only where their cones fall or where animals carry them). Many species will be unable to keep up with the unprecedented rate of change predicted for coming decades. **Wildlife is dislocated.** Natural communities will fragment as their species adapt in various ways; sensitive species may not survive the transition.[23]

Forests at the drier and hotter edges of their ranges—such as lowland ponderosa pine forests, many interior Douglas-fir forests, and the mixed-conifer forests of southwestern Oregon and northern California—are likely to be replaced by grassland and shrubby woodlands. Overall, the total area of forest in the Northwest will probably shrink as forests migrate upslope more slowly than they are destroyed by disturbances at low-elevation sites. In addition, because mountains are smaller at the top than at the base, species moving upslope will see their habitats contract or even disappear.[24]

With a 4.5° F (2.5° C) warming, sagebrush would replace broad expanses of forest in the central Oregon Cascades. Forests on the range's mostly wooded eastern slopes would shrink nearly in half, while sagebrush steppe—not found in the area at present—would cover fully half the land. On the western slopes, the productive and commercially valuable western hemlock zone would shrink by a third or more. One study predicts that western hemlock would disappear completely from the Oregon Cascades.[25]

The same warming would reduce alpine habitats by three-fourths on the central Oregon Cascades' eastern slopes and eliminate them entirely from the western slopes. Trees are already encroaching upon alpine meadows in the Olympics and Cascades; this trend is likely to accelerate and spread with climate change, threatening both rare alpine plants and recreational activities. Nearly half of Mount Rainier National Park's two million annual visitors, for example, go to the summer wildflower displays at Paradise.[26] **Forests and alpine meadows shrink.**

Transitions to new vegetation types will not be gradual affairs driven by slowly changing average conditions. They will likely be triggered by sudden catastrophic disturbances. Such disturbances will probably have much greater impacts on forests than the direct effects of long-term heat or moisture stress. Mature trees themselves endure many swings in climate over their lifespans and can withstand years of drought, but they are much less able to withstand the fires, pest outbreaks, or other disturbances that may come with climate changes. Several costly forest pests—such as budworms in coastal Alaska and woolly aphids in the Cascades—are limited now by cold temperatures. As the Northwest warms, however, their outbreaks will likely grow larger and more severe. In large areas of the B.C. interior, for example, warm winters in the late 1970s and early 1980s allowed unusually large numbers of overwintering bark beetles to survive; the region's lodgepole pine forests suffered massive bark beetle infestations as a result.[27]

As snowpack melts earlier in the spring and higher temperatures cause water to evaporate more quickly, forests will become drier during the Northwest's already dry summers. Forest fires will burn more frequently and probably more intensely. The risk of catastrophic fires in the central Washington Cascades, for example, could triple (from an average of one major fire every 425 years to one every 114 to 166 years). Fires could assume a major role in wet spots like southeast Alaska, which have so far had few. Fires could also cause some sites to shift from conifers to hardwood species like alder and madrone (arbutus), which reproduce readily after burning.[28] **Fires and pest outbreaks intensify.**

In the rain forests of the Northwest coast, wind, not fire, is the major agent of forest disturbance. Global warming will undoubtedly change the world's wind patterns, though how they will change is highly uncertain. Winds are driven by differences in temperature, so they might diminish as the Arctic warms more than regions to the south. Conversely, winds blowing off the Pacific might grow stronger as continents heat up more than the oceans. The number of days with gale-force winds in southeast Alaska has more than doubled over the past 20 years—a period of pronounced warming there. But right now, it's impossible to predict whether the risk of major windstorms—and tree falls—on the Northwest coast will increase or decrease with climate change. Either way, the patterns of disturbance to which native plant and animal species are intimately adapted are going to change.[29]

Rapid climate change will further deplete the region's already scarce reserve of old-growth forests. Old-growth forests, of course, cannot migrate or disperse on any human time scale. Instead, centuries-old forests will be replaced by tiny seedlings that, if they survive, will take centuries to assume the grandeur and ecological value of what was lost. And unlike old trees, seedlings are very sensitive to summer drought and are less likely to survive in the altered conditions.

Since higher temperatures mean longer growing seasons, tree growth may actually benefit from climate change in some places. Where water availability is not a problem, the productivity of Northwest forestry and agriculture could get a boost. Some areas of only marginal value for agriculture today may be able to produce higher-value crops. But one-third of Northwest cropland depends on irrigation for its productivity, and much of the rest may require irrigation if summer precipitation or soil moisture declines.[30]

In addition, the fertilizing effect of carbon dioxide may, in itself, be a boon to plant growth. **CO_2 fertilizes plants.** But it also has a downside. Cheatgrass, the widespread Eurasian invader of grasslands throughout the interior Northwest, could be given a double boost by climate change. More frequent fires would help cheatgrass spread, and like many weedy species, cheatgrass increases its growth rates in CO_2-enriched air more than other plants do. Increased CO_2 could also decrease the nutritional value of vegetation (as plant cells come to contain relatively more carbon and less nitrogen). The ecological effects would be wide-ranging: if plant-eating insects have to eat more to subsist, they will probably grow and reproduce more slowly, reducing the food base for insect-eating songbirds and the birds' predators as well.[31]

YOU WOULD BE HARD-PRESSED to describe Olympia's Port Peninsula—with its sewage treatment plant, container cranes, and vacant lots—as the city's most scenic locale. Much of it has the blighted feel that comes with underused urban land, from the abandoned and toxic creosote plant to the deserted Sea Mart mini-mall. At low tide, nearby mudflats combine with a Kentucky Fried Chicken outlet to give the peninsula's western edge an indescribable aroma.

Yet the port area of Washington's low-lying capital city has its charms. A new farmer's market, a new boardwalk park, and new restaurants are breathing life into the peninsula. It certainly doesn't deserve what's going to be heaped upon it: acres of seawater. Olympia sits on a southern arm of Puget Sound, and much of its downtown and port area was claimed from the sea. About 30 city blocks were built on landfill dredged from Budd Inlet; they lie 4 to 8 feet above ordinary high tide. Much of this area, already prone to flooding during storm surges and unusually high tides, could be reclaimed by Puget Sound with global warming.

When water is heated, it expands. Over the past century, as the world's oceans warmed and glaciers melted, global sea level rose an average of 4 to 10 inches (10 to 25 centimeters). Because of the greenhouse gases already in the atmosphere, the seas will further swell an estimated 3 inches (7 centimeters) by 2020 and 8 inches (20 centimeters) by 2050. Sea level will rise faster in the coming century as more greenhouse gases are added to the atmosphere. IPCC's best guess is that seas will have risen 20 inches (50 centimeters) by the year 2100 if the world fails to control its CO_2 emissions. **Sea level rises.** These amounts may sound insignificant, but even small rises can cause serious erosion. On sandy shorelines, rising seas' effects may be magnified 100 times horizontally: a 3-inch rise can erode 300 inches (25 feet) of beach.[32]

Sea-level rise varies greatly from place to place, especially along the geologically complex and active west coast of North America. Along most of the Northwest coast, plate tectonics, glacial retreat, and other geological factors are uplifting the land and muting the effects of the sea's global expansion. Where land is subsiding, as in the Fraser River delta, southern Puget Sound, and the central Oregon coast near Tillamook, coastal communities and ecosystems will be more vulnerable to rising seas (see Figure 1). Tacoma is subsiding fastest of all, but the city generally sits higher than Olympia and is at less risk than its southerly neighbor. Overall, the Northwest coast is subsiding more slowly and has steeper shorelines than many Atlantic and Gulf Coast locations, which will probably feel the effects of rising seas sooner and more severely than the Northwest.[33]

Olympia is the city most likely to lose prime downtown real estate, but it is not the only place that will be affected. A 12-inch (30-centimeter) sea-level rise would inundate sections of Highway 101 on the Oregon coast. A 20-inch (50-centimeter) rise would permanently flood 45 percent of Washington's tidal mudflats, eliminating critical

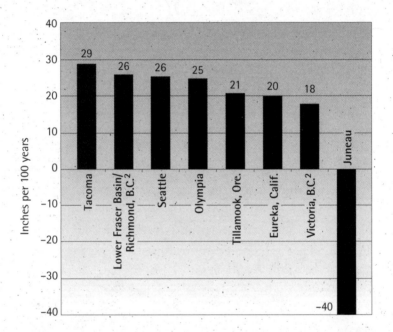

Figure 1. Predicted Rise in Sea Level, Pacific Northwest, Twenty-First Century[1]
The sea will rise fastest in southern Puget Sound, where land is subsiding.

[1] Relative to local land movement; assumes IPCC-estimated global rise in sea level of 20 inches per century, or 5 millimeters per year. One inch equals 25.4 millimeters.
[2] Midpoint of 4-inch range.
Sources: see endnote 33.

habitat for shorebirds and shellfish. **Coastal areas flood and erode.** Rising seas would also cause widespread flooding in Richmond, Langley, and other areas of the Fraser River delta, B.C.'s richest agricultural area. Upgrading the dikes that protect the delta's farms and homes from tidewaters could cost hundreds of millions of dollars, yet would do nothing to prevent saltwater intrusion of wells. Such shoreline defenses also worsen the loss of coastal wetlands by stopping them from migrating inland with rising seas.[34]

Because water takes much longer than air to heat up, changes in sea level generally lag years or decades behind global climate changes. As with other effects of global warming, extreme events are more worrisome than slowly changing average conditions. Rising seas are most likely to take a toll when they combine with other factors—such as storms, rainfall, and high tides—that increase the risk of coastal flooding. Similarly, storms and waves are able to surge farther inland and do more damage when the sea is higher: a 20-inch increase in sea level would give a 75-year storm as much damaging punch as a 100-year storm has now.[35]

Seasonal winds and currents can swell seas temporarily. Tides are typically 4 to 8 inches (10 to 20 centimeters) higher in winter along the Northwest coast because of currents pushing water toward shore. Storms blowing ashore can also raise tides several inches. El Niño—the periodic surge of warm water in the tropical Pacific that plays havoc with weather worldwide—can even raise sea level as far north as Canada. In extreme El Niño years, warm water can surge all the way from Ecuador to central British Columbia, raising sea level along the way. The 1982–83 El Niño raised B.C.'s sea level 8 inches for the entire winter and caused flooding at the mouth of the Fraser River as well as serious erosion along the Oregon coast.[36]

El Niño events have occurred more often over the past 20 years. The El Niño of 1990–95 was the most prolonged on record, and the 1997–98 El Niño is expected to be the strongest this century. If global warming is increasing the frequency or intensity of El Niños—as some scientists suspect—the impacts on coastal zones will be doubly powerful. Even if El Niños do not intensify, the Northwest will probably feel the brunt of a rising Pacific most acutely during El Niño winters. Think of the "super El Niño" winter of 1997–98 as a dress rehearsal for the full force of climate change.[37]

THE PACIFIC NORTHWEST is one of the world's wealthiest and most technologically advanced regions. Endangered species and sensitive ecosystems may have little hope if our climate changes rapidly, yet our economy may be able to absorb some of the impacts of climate change. It is conceivable that the region could muster the billions of dollars necessary to continually fend off rising seas, develop new crops, repair flood damage, and make up for lost hydropower. At the very least, we have the ability to lessen the suffering of our society's most vulnerable members.

But most of the world is far too poor to consider such options. Impoverished farmers cannot build high-tech irrigation systems or buy drought-resistant seeds. Residents of small island nations have nowhere to hide from rising seas. Worldwide, about 46 million people a year already face coastal flooding due to storm surges; a 20-inch (50-centimeter) sea-level rise would double this number. Millions of coastal residents and farmers in drought-stricken areas could become greenhouse refugees, forced to flee their homelands in search of a safer climate.[38]

George Bernard Shaw once said that "one mark of the educated person is that he [or she] can be emotionally moved by statistics." I suppose that makes me an educated person. Knowing how many acres of forest could be lost or how much of Olympia could be swamped by rising seas can move me to sadness and anger. But beyond the regional statistics I've immersed myself in to write this book, I think of my friends where I used to live, in Indonesia. Where will the small farmers on crowded Java go when they can no longer count on the rains to keep their rice paddies flooded? What will the fishing families of the Togian Islands do when warmer waters kill the fish-studded coral reefs or when rising waves flood their sea-level villages?

With the Northwest economy increasingly connected to those on the rest of the Pacific Rim, climatic impacts elsewhere in the world are sure to reverberate here. As Raul Estrada-Oyuela, the Argentine chairman of the global climate treaty negotiations, said, "We are all adrift in the same boat. And there's no way that only half the boat is going to sink."[39]

What Goes Up

Emissions

To WRITE ABOUT snowline on Mount Hood, I needed to see it for myself. But I didn't relish the thought of sitting on I-5 for hours in Labor Day weekend traffic, and I didn't really want to burn a whole tank of gasoline just to write a two-page passage in a book about climate change.

Gasoline is about 85 percent carbon by weight. Two inevitable products result from burning a carbon-bearing fuel such as gasoline: energy and carbon dioxide. Every gallon I burn in my engine sends about 20 pounds of carbon dioxide, containing 5 pounds of carbon, into the atmosphere. It's like tossing a 5-pound bag of charcoal briquettes out my window every 20 miles or so. (A Canadian driver sends 0.6 kilograms of carbon [2.4 kilograms of CO_2] into the air for every liter of gasoline.) Include the fuel burned in drilling, refining, and transporting the gasoline, and those CO_2 emissions go up another 20 percent.[40]

Carbon dioxide from burning fossil fuels is the primary cause of global warming, and here in the Northwest, cars are the number one source of carbon dioxide. With a little easy math, I calculated my personal impact: at 25 miles per gallon, my car trip from Seattle to Mount Hood would boost the Northwest's contribution to climate change by 420 pounds (190 kilograms) of carbon dioxide—more than twice my own weight.

IN 1992, THE LARGEST assembly of national leaders in history convened in Rio de Janeiro at the Earth Summit. The heads of state agreed that the Earth's climate was seriously threatened and committed to take action to protect it. They signed the Framework Convention on Climate Change, whose ultimate goal is "stabilization of greenhouse gas concentrations in the atmosphere at a level that would prevent dangerous anthropogenic interference in the climate system." To achieve this goal, the world as a whole will need to slash its greenhouse gas emissions. As a first step, industrial nations—which produce the vast majority of atmospheric pollution—agreed to cap their emissions of greenhouse gases at 1990 levels by the year 2000.

Five years later, few nations are within reach of that goal. Most have let their emissions rise, some dramatically. The United States, the world's largest polluter of the atmosphere, increased its carbon dioxide emissions 8.7 percent from 1990 to 1996. Canada, which uses even more energy per capita than the United States, watched its greenhouse gas emissions grow 9.2 percent from 1990 to 1995, faster than the nation's economy or population. If current trends continue, both nations will overshoot the climate treaty's goals for the year 2000 by 15 to 20 percent.[41]

North America doesn't lead the world in greenhouse gas emissions because its people are greedy or enjoy polluting. Greenhouse gases are invisible, their impacts easy to forget. Low energy prices enable most of us to get on with our busy lives without worrying about how much energy we use or how much invisible pollution we create. And policies from building codes to transportation funding discourage us from living with less environmental impact.

The trickiest thing about greenhouse gases is that they're everywhere. There's no one dike we can put a finger in to stop the flood of heat-trapping substances—including carbon dioxide, methane, nitrous oxide, perfluorocarbons, and others—from rushing into the atmosphere. These gases come not just from belching factories or tailpipes but from the most mundane of activities. Turning on a light, eating lunch, reading the paper, dumping the trash—all have an impact on the climate.

Of all the greenhouse gases, carbon dioxide is by far the most important, and fossil fuel burning is by far its most important source. The Pacific Northwest's emissions of CO_2 from fossil fuels have grown rapidly, by 9 percent from 1990 to 1994 and by more than a quarter over the past decade (see Figure 2). Emissions rose in every part of the region in those four years: 16 percent in Oregon, 14 percent in Idaho, 7 percent in Washington, and 6 percent in British Columbia. B.C., which has done a more comprehensive accounting of its atmospheric pollution than any other jurisdiction in the Northwest, registered 16 percent growth in overall greenhouse gas emissions from 1990 to 1995. Without major policy changes, emissions from every jurisdiction in the Pacific Northwest will keep rising until the year 2000 and beyond.[42]

Since 1960, when detailed record keeping of fossil fuel consumption began, CO_2 emissions from these fuels have

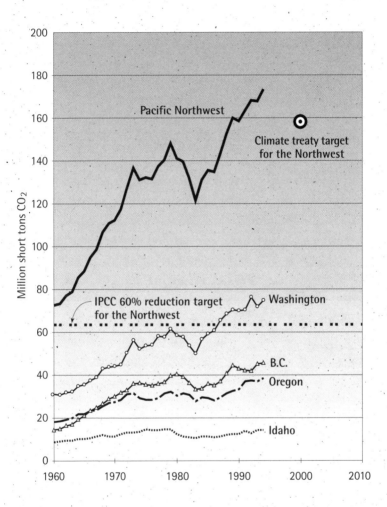

Figure 2. Emissions of CO_2 from Fossil Fuels, Pacific Northwest, 1960-94
Northwest emissions are overshooting the targets for climate protection.
Sources: see endnote 42.

more than doubled in Oregon, Washington, and the Pacific Northwest as a whole. In Idaho, they have increased by two-thirds; in British Columbia they have tripled.

Why has our impact on the atmosphere grown so dramatically? As our economies and populations have grown, so has our use of petroleum, natural gas, and coal. Every sector of the economy consumes fossil fuels, but transportation is the main culprit. Industrial and residential heating and electricity generation also use large amounts. Gasoline and diesel fuel are the source of 47 percent of the region's fossil fuel–based emissions; natural gas (both for heating and for generating electricity) is the next largest at 25 percent. Coal (10 percent), jet fuel (9 percent), and fuel oils favored by industry (8 percent) are also major sources.

The Northwest's emissions are higher in the 1990s than ever before, but per capita, our rates actually peaked 25 years ago (see Figure 3). Our energy use and our economy have not grown in lockstep. Rapid economic growth and inefficient energy use in the 1960s combined to send the region's fossil fuel consumption and pollution upward fast. But the 1973 OPEC oil embargo, which raised gas prices suddenly and sharply, brought a quick end to the days of carefree energy use. Higher energy prices from 1973 on induced people to consume less; government policies promoting energy efficiency and conservation helped people do more with less.

Emissions also fell in the 1970s as natural gas replaced petroleum as the region's main source of heat (after the sun, of course). The switch was most striking in B.C., which tapped into the gas fields in the province's northeast and in neighboring Alberta. British Columbia today gets more

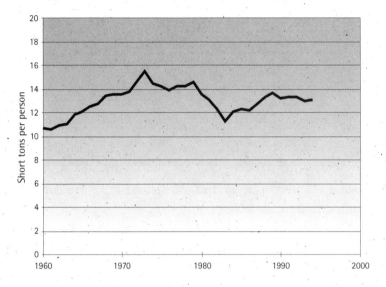

Figure 3. Per Capita Emissions of CO_2 from Fossil Fuels, Pacific Northwest, 1960–94
Pollution drops when energy prices rise, as from 1973 to 1983.
Sources: see endnote 42.

of its energy from natural gas than from any other source. Natural gas burns much cleaner than oil or coal, releasing one-sixth to one-third less carbon dioxide per unit of energy than petroleum fuels, and half as much as coal. But it is less "clean" than its advertisements make it out to be. Natural gas is mostly methane—a far more potent greenhouse gas, molecule for molecule, than carbon dioxide. Raw natural gas also contains an average of 7 percent carbon dioxide. Leaks from wells and pipelines and carbon dioxide discharges in processing boost greenhouse gas emissions from natural gas. And clearing boreal forests in the Canadian Rockies to explore and drill for natural gas releases still more carbon dioxide into the atmosphere. Nonetheless, natural gas remains the fossil fuel least damaging to the global climate.[43]

I DECIDED TO TAKE the train to Oregon. Instead of staring at taillights for eight hours, I'd be able to relax. I'd also do less damage to the climate. From Portland's Amtrak station, I'd rent a car and drive the short last leg to the trailhead on Mount Hood.

As one of several hundred passengers on the train, I was responsible for a fraction of the diesel fuel it burned going down the tracks. My share from Seattle to Portland was about 160 pounds (72 kilograms) of carbon dioxide, just under half the pollution I would have caused had I driven. The driving leg from Portland up the mountain and back was the same either way. So by taking the train, I saved my own weight in greenhouse gases.

ACCORDING TO THE U.N. Intergovernmental Panel on Climate Change, an immediate 60 percent reduction in global emissions is necessary just to stabilize concentrations of CO_2 in the air at their present elevated levels. Industrialized nations would need to reduce their emissions even more to stabilize the global atmosphere while allowing poor nations reasonable economic growth. With many North American policymakers actively fighting even the most tepid attempts to cap emissions at their current levels, reducing them by more than 60 percent may seem an outlandish goal. It's the sort of proposal legislators would dismiss as "a nonstarter." But unlike legislation, the laws of atmospheric physics can't be revised in response to focus groups or weakened to please campaign contributors. The science unequivocally demonstrates our need to reduce global emissions as quickly as possible to fight global warming. Fortunately, a quick look at the history of emissions in the Northwest gives cause for hope: rapid reductions are within our reach.

For most of the past three decades, the Northwest has been growing more carbon efficient: the economy now generates less carbon dioxide for every dollar it produces. The region was especially successful at economizing on fossil fuels during the energy-conscious 1970s. In the ten years following the OPEC oil embargo, the region's total CO_2 emissions fell 10 percent, even as the economy grew 44 percent. Per capita emissions dropped from their 1973 peak of 84 pounds (38 kilograms) per person per day to 61 pounds (28 kilograms).[44]

Then, in the 1980s, energy prices declined, as did government support for programs to improve energy efficiency, and the progress of the 1970s stopped. Our economy generated as much fossil fuel pollution per dollar in 1993 as it had ten years earlier. Per capita CO_2 releases, meanwhile, rebounded from their 1983 low to the gas-guzzling levels of the early seventies, and total emissions reached record-breaking levels. The 1980s were a lost decade for climate protection.

Now look again at Figure 2 (page 35) and imagine a different scenario. What if, instead of abandoning energy conservation, importing more oil, and letting CO_2 emissions skyrocket, the Northwest had continued economizing on carbon dioxide? What if the Northwest kept shaving the emissions for every dollar of wealth generated as it had from 1973 to 1983? By 1994, the region's total emissions would have fallen to pre-1970 levels, 25 percent below their 1979 peak. What's really impressive is that a 25 percent reduction in CO_2 emissions in only 15 years would have been more dramatic than anything proposed at the 1997 global climate treaty talks. The Northwest would be well on its way toward building an economy that behaves responsibly toward the climate. Now imagine what we could achieve if we really set our minds to it.

IF I HAD MADE my trip to Mount Hood earlier in the summer, my life would have been a lot easier. Snow would have covered more of the mountain, and I would have had a shorter hike to snowline. As it turned out, I had to scramble off-trail for several hours, into and out of the White River Canyon, to reach snow.

Hiking longer than I expected, I ended up running down the canyon trail to reach the car in time to drive back to Portland and catch the last train of the day home. The road at Wapanitia Pass was being repaved, and I was delayed even more. I drove like a madman down Highway 26 until hitting Portland rush-hour traffic. I arrived at Union Station half an hour late and jumped aboard just before the delayed train pulled away.

It may have been an act of God that got me on the train, but my driving 80 miles per hour in a 55-mph zone undoubtedly helped. Driving so fast also boosted my small car's fuel consumption by about 30 percent—making it barely as efficient as a sport utility vehicle nearly twice its weight. Add another gallon of gas and another 20 pounds of CO_2 to my trip report. Toss another bag of briquettes on the global barbecue.[45]

CARS AND TRUCKS are the Northwest's largest source of greenhouse gases. More drivers, each driving more miles, have been a recipe for surging emissions. Vehicles outnumber registered drivers in the Northwest, and each of those vehicles is, on average, driven farther than ever before. In British Columbia, for example, the distance driven per capita doubled between 1970 and 1995; the total distance driven more than tripled.[46]

The main reason for this increasing auto dependence is the suburbanization of the Pacific Northwest. Suburbs have been sprawling across forests and farmland and forcing people who live or work there to use their cars more. With amenities scattered and landscapes hostile to any form of transportation except driving, suburbs effectively lock their residents into heavy car use—and carbon dioxide emission. Suburbs are now home to more of the region's people than are cities, towns, or rural areas.[47]

Over most of the past 30 years, gasoline consumption grew more slowly than miles traveled because cars were becoming more fuel efficient. But that trend has been reversed in the 1990s—the decade of the sport utility vehicle (SUV). The fuel efficiency of the American automobile fleet peaked in 1988 (shortly after oil prices crashed) and has been stagnant or falling ever since. Oil prices have remained at historic lows since the mid-eighties: taking inflation into account, U.S. gasoline in 1997 is little more than half the price it was in 1980. With their unusually low gasoline taxes, the United States and Canada are among the few places in the world where a gallon of gas costs less than a gallon of bottled water.[48]

Spurred by low fuel prices in the 1990s, North American sales of "light trucks" (the category including sport utility vehicles, minivans, and pickups) have risen much faster than car sales. American SUV sales have tripled in the ten years since 1986. By 1995, 39 percent of all new passenger vehicles sold in the United States—and 44 percent of those sold in Canada—were light trucks. Currently, about 29 percent of all "cars" on North American roads are actually light trucks; light truck sales are expected to overtake new car sales before the year 2000.[49]

The surging popularity of SUVs is bad news for the climate. They are the least efficient of all major forms of transportation (see Figure 4). A new SUV goes through gasoline at least a third faster than a new car. Under the CAFE ("corporate average fuel economy") standards, new cars sold each year in the United States must average 27.5 miles per gallon; light trucks are required to average only 20.7 mpg. Even those figures are artificially high: vehicles are almost always less efficient than their official ratings, and sport utility vehicles are less efficient than other light trucks. On the road SUVs actually average about 15 miles per gallon. (Take an SUV off-road in four-wheel drive, and it averages only 10 miles per gallon.)[50]

All indications are that sport "ute" sales will continue to grow as long as gasoline remains cheap and CAFE standards remain as low as they are. Detroit's Big Three auto-

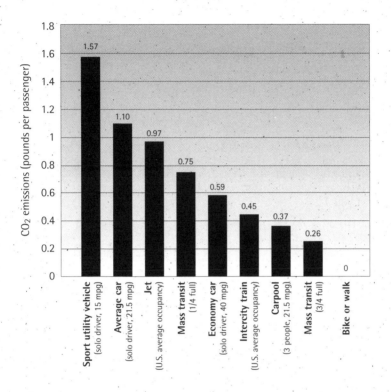

Figure 4. CO_2 Emissions per Mile Traveled
Taking a bike, bus, or train pollutes less than driving alone.
Sources: see endnote 50.

makers earn all their profit on light trucks; their cars are actually money losers. The vehicles themselves are exempt from the "gas guzzler" tax applied to large cars. A new generation of even larger sport utility vehicles is on the drawing boards in Detroit, portending more traffic fatalities among drivers of regular-sized cars and more damage to the climate.[51]

ON THE TRAIN ride home, I decompressed from my frenzied trek down from Mount Hood and watched the Columbia River roll by. We passed the world's largest egg (it's fiberglass) on proud display in the tiny hamlet of Winlock, Washington, and not long afterward stopped in Centralia.

Centralia is a sleepy town in the cutover lowlands of southwestern Washington. It happens to be home to the single largest disrupter of the global climate in the Pacific Northwest. About 5 miles northeast of town lies the Centralia Coal Mine and, next to it, the Centralia Steam Plant, the Northwest's largest coal-burning power plant.

Coal, the most abundant and carbon intensive of all fossil fuels, is generally little used in the Northwest. Most electricity in the region comes from dammed rivers, not fossil fuels. Because of this, the region's per capita carbon dioxide emissions are considerably lower than for the United States or Canada as a whole.

In a region dominated by hydroelectricity, Centralia is a highly polluting anomaly. The local coal it burns is the dirtiest in the West—emitting more carbon dioxide, and more health-threatening sulfur oxides, than any other major fuel. Centralia emits 2.6 pounds (1.2 kilograms) of CO_2 per kilowatt-hour of electricity it generates—three times what today's natural gas–burning power plants release. In 1994, this one plant was responsible for 15 percent of all carbon dioxide emissions from fossil fuels in the state of Washington. It emitted as much carbon dioxide as 2.5 million cars—roughly half of those on Washington's roads.[52]

In the United States, coal provides 56 percent of the nation's electricity supply, natural gas another 10 percent.

Coal and other fossil fuels provide 19 percent of Canada's electricity. Because much of the continent is wired together by an electric grid, with power zapped back and forth on demand, electricity use in the Pacific Northwest affects power production elsewhere. In the Northwest, as elsewhere in North America, hydroelectric and nuclear power generation have probably peaked: the best rivers have been dammed, and nuclear power remains too expensive and risky for new plants to be built. If anything, hydropower generation will probably shrink as dam operations are adjusted to lessen their impacts on salmon and rivers and as the Columbia River's flow declines with climate change.

For these reasons, though northwesterners may think we have clean electricity that does not affect the climate, it does. At the margin, a kilowatt-hour used here probably means more fossil fuels burned here or elsewhere in western North America. And a kilowatt-hour conserved is, most likely, fossil fuels not burned.

Electric utilities, like phone companies a decade ago, are in the midst of a profound transformation: from locally regulated monopolies to competitors on an open market. Every state and province in the Northwest either has deregulated or is considering deregulating its utilities. This means that, before too long, I'll be able to choose what kind of electricity I use—and that I'll soon be hit with a flood of solicitations from power companies like the ones I get now from long-distance phone companies.

Since the 1980s, new sources of natural gas and new gas-fired technologies have driven down the cost of electricity, and industries have pressured governments to open markets to the cheaper electricity. Facing increased competitive pressure, utilities in the Northwest are cutting costs. Bonneville Power Administration, the Northwest's largest electricity provider, slashed its conservation budget 92 percent between 1994 and 1997. Other utilities in the region have cut their funding for energy efficiency and wind and solar power by more than half. At a time when electricity demand is rising and dams need to let more water bypass their turbines to save endangered salmon, these spending cuts translate directly into more fossil fuel emissions.[53]

Homes use about a third of all the electricity consumed in the Northwest, with business and industry accounting for the other two-thirds. One reason that consumption of electricity (and natural gas) is on the rise is that Americans are, on average, living in bigger houses. Even as the number of occupants per household has been dropping, the size of the average new American house has been growing: from 1,600 square feet (150 square meters) in 1975 to 2,100 square feet (195 square meters) in 1995. "Everybody wants four bedrooms when they only use two for sleeping," according to Gopal Ahluwalia of the National Association of Home Builders. Larger houses also mean more construction materials and more greenhouse gas emissions from cement and other factories. (Carbon-rich limestone is the main raw material, and carbon dioxide a major by-product, of cement manufacturing.)[54]

Concerned about the environmental impact of home construction and use, a number of architects, builders, and home buyers in the Northwest have turned to energy-efficient appliances and materials such as compact fluorescent light bulbs and superinsulated windows and walls.

These efforts deserve to be redoubled, but they should be kept in perspective, too. A study in Portland found that a family living in a drafty, 80-year-old energy hog of a house in a traditional city neighborhood will still use less energy than a family living in a highly efficient new home in the suburbs. Despite reduced electricity and gas bills in the new suburban home, total energy use is greater for the suburban family because they end up driving three times as much as they would in the city. A potential home buyer would help the climate by choosing an urban home and lifestyle over a suburban one. *Where* homes are built matters even more than *how* they are built.[55]

ON A CLEAR DAY, stand atop Mount Hood or almost any other mountain in the Cascades or the Coast Range. You will be struck by both the scenery and the clearcuts that have become as much a part of the Northwest landscape as evergreen trees themselves. The forests in this part of the world have been heavily cut because they are exceptionally heavy with wood. The Northwest's coastal rain forests are champions among ecosystems at putting on weight: more biomass is stored in an acre of trees, fallen logs, and soils of our coastal rain forests than in any other acre on Earth.[56]

The basic building block of living tissue, of biomass, is carbon. When that tissue decays, carbon returns to the atmosphere as carbon dioxide (or as methane if it decays in the absence of oxygen in someplace like the bottom of a landfill). Over the past century, converting the ancient coastal forests of the Northwest to younger stands of smaller trees has added billions of tons of CO_2 to the atmosphere. Some of the forests' carbon is stored for decades in two-by-fours, furniture, and other durable wood products. Yet burning and decay—of wood and bark left in the forest, chips and sawdust created in mills, and paper products—rapidly release most of the carbon into the air. Forest regrowth does take up some of this carbon dioxide over later decades; but second-growth forests are usually cut more than a century before they can recapture all the carbon lost from their old-growth predecessors. The net effect is to add carbon dioxide to the atmosphere.[57]

By the 1990s, relatively little old growth remained in Oregon or Washington, and logging there released relatively little carbon into the atmosphere. In British Columbia and southeast Alaska, however, the timber industry continues to rely almost exclusively on old-growth forests for its raw material. The industry continues to be one of British Columbia's largest sources of greenhouse gas emissions. Fully one-fourth of the province's contribution to global warming in 1994 came from cutting its coastal rain forests (see Figure 5).[58]

Coastal logging will eventually decline in B.C., as it has to the province's south. The finite supply of accessible ancient forest will sooner or later give out—but maybe not before millions of additional tons of greenhouse gases are emitted and other irreversible ecological damage is done.

THE INVISIBLE SURGE of heat-trapping gases rising from the Pacific Northwest is mostly carbon dioxide, but three lesser ingredients boost its potency. These gases—

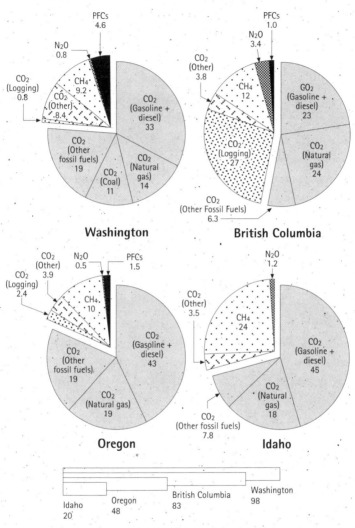

Figure 5. Greenhouse Gas Emissions, 1994
Gasoline and other fossil fuels dominate our atmospheric pollution.
Sources: see endnote 58.

methane (CH_4), perfluorocarbons (PFCs), and nitrous oxide N_2O)—hold warmth scores, even thousands, of times more effectively than carbon dioxide. Less than 1 percent by weight of the region's greenhouse gases, these chemicals cause 15 percent of our damage to the global climate.

A tenth of the Northwest's contribution to global warming in 1994 came from methane. Its main sources weren't pretty: decomposing garbage, the guts and manure of livestock, and natural gas leaks. In Idaho—a state where cows outnumber people—methane caused a fourth of the state's contribution to climate change. Methane emissions from U.S. landfills have dropped dramatically since the U.S. Environmental Protection Agency (EPA) began requiring large landfills to capture more of their gas and flare it or sell it for fuel. Either way, the methane is converted to carbon dioxide—which is only 5 percent as powerful at trapping heat as methane. If sold as fuel, the landfill gas further serves to displace fossil fuel combustion. Methane emissions in B.C., by contrast, are on the increase: most of the methane from landfills still escapes to the atmosphere, and the province's cattle population and natural gas production are both rising.[59]

Perfluorocarbons come from only one source in the Northwest: aluminum smelters. Though smelters produce these chemicals in very small amounts, PFCs are 6,000 to 9,000 times more powerful than carbon dioxide at trapping heat. They are also the longest-lived of all greenhouse gases: if the first North Americans who migrated across the Bering land bridge 30,000 years ago had been able to create the PFC perfluoromethane, much of it would still be trapping heat in the atmosphere today. PFCs make aluminum smelting one of the most climate-threatening industries known. Washington's seven aluminum smelters caused nearly 7 percent of the state's contribution to global warming in 1994.[60]

A smelter uses large quantities of electricity to jolt oxygen atoms away from alumina (aluminum oxide) and attach them to electrodes made of carbon, forming pure aluminum and carbon dioxide. Smelters produce nearly twice as much carbon dioxide as they do aluminum. When the concentration of alumina in a smelting pot drops too low, PFCs are produced. On average (since smelter technologies vary greatly), a ton of aluminum smelted means the greenhouse equivalent of 4 to 5 tons of carbon dioxide sent into the atmosphere.[61]

All ten aluminum smelters in the U.S. Northwest have joined the EPA's Voluntary Aluminum Industrial Partnership, a program started in 1995 aimed at reducing PFC emissions by 45 percent by the year 2000. The EPA estimates that participating smelters have reduced their PFC emission rates by 30 percent to date. The Alcan smelter at Kitimat, B.C., reports a 10 percent reduction in PFC emissions in the 1990s. Major reductions in carbon dioxide emissions are not expected at any smelters, barring major technological breakthroughs that would replace the use of carbon electrodes.[62]

ALL TOLD, the Northwest's greenhouse gas emissions are equivalent to more than 100 pounds (45 kilograms) of carbon dioxide per person per day across the region—some of the world's highest rates of pollution (see Table 1). The

Table 1. Per Capita Greenhouse Gas Emissions, 1994
Northwesterners are world-class polluters.

Region	CO_2 from fossil fuels (tons[1])	All greenhouse gases (tons CO_2 equivalent)
British Columbia	12.5	22.6
Idaho	12.5	17.6
Oregon	12.5	15.4
Washington	14.0	18.4
Pacific Northwest	**13.1**	**18.8**
Canada	16.9	22.6
United States	21.5	25.9
Western Europe	7.6	n.a.
Africa	1.2	n.a.
World	**4.4**	**6.2**
IPCC 60% reduction target[2]	1.8	n.a.

[1] *One ton equals 0.907 metric ton.* [2] *At 1994 world population.*
Sources: see endnote 63.

Pacific Northwest vents almost 40 percent less CO_2 from fossil fuels than the U.S. average, and one-fifth less than Canada's average, because we get most of our electricity from dams rather than coal. Compared with the rest of the continent, the Pacific Northwest looks good, but by any other standard, we are world-class polluters. Our per capita CO_2 emissions are nearly twice those of western Europe, where economies are highly energy efficient; they are three times the global average.[63]

When greenhouse gas sources other than fossil fuels are added in, the gap between the Northwest and the rest of North America narrows. Cheap hydropower has made the Northwest a center of climate-damaging aluminum smelting, and depletion of coastal rain forests boosts the region's greenhouse tab. British Columbia, with huge areas of rain forest being converted to tree farms, has the region's greatest per capita greenhouse gas emissions—more than four times the global average.

Most of the world aspires to live like North Americans, but this would be a recipe for climate catastrophe. The world already emits more than twice as much carbon dioxide as it can if we want to stabilize our atmosphere. In other words, we'd need another whole planet's worth of air to safely absorb all the pollution we create. If everyone on Earth lived like northwesterners, the planet's greenhouse gas emissions would triple; our climate would quickly unravel. The world's emissions would be seven times the atmosphere-stabilizing level. In other words, we'd be at least six atmospheres short of a stable climate.

It's up to us in the highly polluting minority to take the lead and show the world there are better ways to live.

What We Can Do

15 Steps to Climate Leadership

WHEN FRIENDS ASK me what I've been up to lately, I tell them I've been finishing my book on climate change. Most ask when I'll be done or chide me for working too much through a sunny Northwest summer. Few ask about the book itself; its topic floats by without registering any emotional response. Even though writing this book meant poring over reams of dry reports and data tables, the topic of climate change has definitely hit me emotionally. When I think about Glacier National Park without glaciers or the Columbia and Fraser Rivers with fewer salmon, I can't help feeling pain and loss. Those feelings keep me working and doing what else I can to minimize my own impact on the atmosphere. ❶ **Visit a favorite place:** a beach, a cousin's farm, a fishing spot, a nature trail. Think about what could happen to it if our climate unravels.

Though the scientific evidence that we need dramatic action to protect our climate is beyond dispute, many North Americans think that the issue is still unresolved, that global warming is still only a theory that scientists argue over. The confusion is not surprising: energy companies and their lobbying groups spend tens of millions of dollars to make sure people think this way. In this era of information overload and slick public relations campaigns, most people stay unmoved or distrustful. They may be moved, however, if those close to them, those whose judgment they trust, share their own knowledge and concern. ❷ **Talk with a friend about the climate.**

Climate change is probably the greatest environmental challenge the world has ever faced, and world leaders are beginning to give it the attention it deserves. Yet individuals don't have to wait for someone else to lead them toward a safer way of living. Each of us has tremendous power to reduce our impact on our world through choices we make every day: transportation we take, products we buy, activities we engage in.

You can figure out how to help the climate by tallying your energy use and other consumption. Transportation puts out more tons of greenhouse gases than any other source in the Northwest, so, chances are, the weightiest decision you will make today is how to get where you're going. ❸ **Do the math.** Calculate the pounds of CO_2 your car emits each day; it's easy. Take the number of miles you drive, multiply by 20, and divide by your car's miles per gallon. (Canadians, multiply kilometers driven by the number of liters it takes your car to drive 100 kilometers, and divide by 42 to get CO_2 emissions in kilograms.) Add another 20 percent to include the CO_2 released in making your gasoline. To do a more comprehensive self-exam, fill out the "Personal CO_2 Calculation" form on the International Council for Local Environmental Initiatives' Web site: *www.iclei.org/iclei/co2calc.htm*.

Even though we need to greatly reduce the impact of our transportation system on the atmosphere, North Americans are speeding in the opposite direction. The latest highway fad—the sport utility vehicle—is the most polluting form of transportation around. ❹ **Ixnay on the SUV.** Let's face it: we don't really need to wrap ourselves in vehicles designed for elephant hunting to hop down to the quickie-mart. Remember: friends don't let friends drive SUVs. If you're buying a new car, go for a fuel miser, and if you have two cars, leave the less fuel-efficient one at home.

It's easy to pin the blame on gas guzzlers and our hopes on "green" cars. But unless we reverse the underlying trend of more vehicles being driven more miles, even the greenest cars using the greenest fuels will consume more energy, cause more accidents, and eat up more money and countryside. Individuals and governments need to support alternatives to driving alone.

For most northwesterners, cars are essential tools of modern life. Yet they are often the wrong tools for the job. Half of all car trips in the United States go 5 miles or less. Try biking or walking to a nearby store instead of driving somewhere far away. For medium-distance trips (such as Vancouver to Seattle or Seattle to Portland), taking the train pollutes much less than driving alone (or flying). And as the Amtrak ads say, on a train you don't have to pull over to kiss. Use your car when you need to, but use something else when you can.[64] ❺ **"Step away from the car, and no one gets hurt!"**

Every hour spent commuting in traffic is an hour less to enjoy your life. Living close to where you need to go reduces stress and increases leisure time. Similarly, everything you buy took energy to make; every dollar you spend took precious time of your life to earn. ❻ **Keep it simple.** Simplifying your life—buying less stuff—is good for the

Consumption and Climate

Northwesterners wanting to protect the climate should have two top priorities: reducing their personal energy use—especially in their cars—and getting involved in the politics of climate change. But watching what you consume can also reduce your greenhouse gas emissions. What you buy makes a difference!

It can be nearly impossible to figure out how much energy went into this widget or that gizmo, let alone products made up of all kinds of parts. But some products clearly have more impact on the climate—virgin materials over recycled goods, for example. As a rule of thumb, consumers can benefit the climate by buying used, recycled, minimally packaged, locally produced, and long-lasting goods. And—as a rule of thumb—avoid things made of virgin metal (aluminum most of all) or old-growth or tropical timber; flown in from across the world (like New Zealand apples or trendy sports shoes); or designed to be thrown out after one use. For major energy users such as appliances, light bulbs, and cars, look for the most energy-efficient model. Even if an energy-efficient good (such as a compact fluorescent light bulb) seems expensive up front, its lower energy costs will likely save you money in the long run.

Consuming smarter is good, but consuming less is probably more important. Simplify your life. Spend less money and time shopping. Spend more time with your kids, or someone else's. They'll be better off, and so will the climate.

climate and good for your happiness (see sidebar). People across the Northwest are aiming to live better lives with fewer things; groups like Seattle's New Road Map Foundation and Portland's Northwest Earth Institute are helping them learn how.

Change often begins with the individual, but the scale of change demanded to stabilize the global climate is so massive that political as well as personal action is required. Swaying policymaking in the right direction is probably the most important individual action you can take.

The politics of climate change is straightforward. By conventional measures, our fossil fuel–driven economy is thriving, and those who profit most from fossil fuel combustion work hard to keep policymakers from acting on the best available science. Without a massive groundswell of action in support of the atmosphere, the divide between science and policy—between knowledge and action—will remain insurmountable. ❼ **Speak out.** Elected representatives, newspapers, and companies all need to hear that our climate is worth protecting. To learn where your efforts can have the biggest impact, contact the groups listed on page 79. Their work deserves your support.

One of the most effective ways to reduce greenhouse gas emissions across the board is to tax them. Widely used in Europe, "green taxes" are effective in steering market forces toward social goals, but they face an uphill political battle in North America. National taxes would be most effective, but if one state or province lowered its income or sales tax and imposed a carbon tax, it might jump-start the process nationwide. Such a tax shift would be doubly beneficial: higher taxes on fossil fuels would discourage their combustion, while lower income taxes would encourage the use of labor—combating unemployment and pollution at the same time.

Carbon dioxide is the most important of the greenhouse gases, but governments should not overlook the other, more potent gases like methane, nitrous oxide, and PFCs. Though they generally have little effect on human health—nitrous oxide is actually used as an anesthetic—these greenhouse gases are dangerous pollutants, and society needs to begin treating them that way.[65] **❽ Tax greenhouse gases.** Instead of making polluters pay, governments are actually paying people to pollute. Dozens of subsidies reward companies and individuals in the Northwest for damaging the atmosphere. In the United States, for example, fuel expenses are tax deductible for utilities and corporate energy users, giving fossil fuels a cost advantage over fuel-free renewable technologies. Aluminum smelters from British Columbia to Oregon depend on subsidized hydropower and emit more greenhouse gases than if they had to pay market prices. Above all others in the Northwest, one set of subsidies is especially egregious: state payments to the Centralia coal plant in Washington, source of 15 percent of the state's CO_2 emissions from fossil fuels. The plant continues to operate only because the equipment needed to reduce its deadly sulfur emissions will be largely paid for by taxpayers. One of the payments takes the form of a special sales tax exemption, worth about $10 million annually, on the dirty coal the plant burns. If this plant were not propped up, market forces would probably shut it down. Closing it would, in one fell swoop, enable Washington to achieve its goal of returning the state's CO_2

emissions to their 1990 levels (even if its power were replaced by natural gas–fired power and not by investments in renewable energy or energy efficiency).[66]

In the United States, simply eliminating subsidies for free parking spaces would reduce CO_2 emissions by 1.2 percent; eliminating highway subsidies would reduce them by 2.0 percent. Ending tax breaks for coal companies and other subsidies in the energy sector could reduce the nation's CO_2 emissions 1 to 4 percent by the year 2010. Furthermore, all these actions would help the economy by saving public funds for truly beneficial purposes or by reducing the tax burden on the rest of society.[67] **❾ Don't pay people to pollute.**

While President Clinton and Vice-President Gore have been voicing concern for the global climate, they have yet to match their words with actions. Like Canada's National Action Program on Climate Change, the Clinton administration's Climate Action Plan relies primarily on voluntary action and has done little to reduce emissions. Over the past few years of negotiations, the main obstacle to an effective global climate treaty has been the refusal by a small group of nations, including Australia, Canada, and the United States, to commit to reductions in CO_2 emissions. Their positions leave North American governments far outside the international mainstream: the *Vancouver Sun* called the Canadian government's efforts to block binding emission limits "a national humiliation."[68]

When leadership is lacking at the national level, it may come at the local level. Kamloops, Vancouver, and ten other local governments in British Columbia have committed to reducing their emissions 20 percent within ten years.

Portland has adopted a goal of reducing its emissions 20 percent below 1988 levels by the year 2010. The Washington cities of Burien, Olympia, and Seattle have made commitments to monitor and reduce their local impact on the global climate, although no results have yet been seen.[69] Just one month before the U.S. Senate unanimously resolved to oppose any climate treaty that required industrial nations to lower their pollution before poor nations did, the Oregon legislature unanimously passed the United States' first law requiring reductions in CO_2 emissions. Any new gas-fired power plants built in Oregon now have to reduce or offset their CO_2 emissions by 17 percent in comparison with the most efficient plant operating in the United States. Similar standards will apply to other new power plants. As better U.S. power plants come on line, new plants in Oregon will face continually tougher standards.[70]

Though this CO_2 law is a small step (it has no effect on existing power plants, for example), it is pathbreaking within its present political context: legislatures across the continent are deregulating the electricity industry and boosting its emissions. Utilities in the Northwest and elsewhere are slashing their investments in conservation and energy efficiency. By doing so, Northwest utilities are abandoning a tradition of leadership in energy conservation: since the Northwest Conservation Act was passed in 1980, the region has saved enough electricity to power a city the size of Seattle.[71] **⑩ Trim the fat:** promote energy efficiency. In an era of global warming, the Northwest should step up, not abandon, its commitment to conserve energy. Legislators in Olympia, for example, should restore funding for the Washington State Energy Office, which was dropped from the state budget in 1996. At a minimum, lawmakers presiding over utility deregulation need to set aside a percentage of regional electricity expenses (what policy wonks call a "system benefits charge") for investments in conservation and efficiency.

Over the past 20 years, federal efficiency standards have cut in half the electricity used by a typical U.S. refrigerator. Strengthened standards could raise the efficiency of a range of products, from dishwashers to cars. ⑪ **Set your standards high.** The U.S. Congress should raise CAFE standards for cars from 27.5 to 45 miles per gallon (from 8.6 to 5.2 liters per 100 kilometers), and comparably for light trucks. (Canada's fuel efficiency standards mirror those of the United States, except that they are voluntary, not mandatory.) Existing technologies can substantially raise fuel efficiency by improving vehicles' aerodynamics, transmissions, and engines. "Hypercars" using cutting-edge materials and technologies can get as much as 300 miles per gallon (0.8 liters per 100 kilometers).[72]

In the long term, the best way to reduce traffic emissions is to reduce the need for travel. City dwellers drive a third as much as suburbanites: when people live close to their jobs, schools, and chores, they drive less. Cities' higher densities—which bring many destinations close enough for walking or biking—also make public transit cost effective. ⑫ **Revive cities and control sprawl.** Read *The Car and the City* (see inside back cover) for more ideas.

Even with vastly improved energy efficiency, the use of renewable energy—such as wind and solar power—will need to grow rapidly to replace declining production from fossil fuels and salmon-killing dams. Despite abundant resources, the Northwest today has no commercial electricity provided by wind or solar power. The Columbia River Gorge is world famous among windsurfers, but its winds remain untapped for their energy. Numerous other sites in the region—including the many steep-sided fjords and inlets along the coast—have powerful winds.[73]

Solar energy may not spring to mind in thinking of the drizzly Northwest, but in southern Idaho, eastern Oregon, and central Washington, solar radiation is almost as strong as in the U.S. Southwest. About 35,000 solar energy collectors have been installed in these three states, the vast majority of them before 1985, when federal tax credits for solar installation expired. Oregon has state tax credits for renewable energy, and solar installation there has dropped less dramatically. Despite fast-falling costs, renewables still need government support to compete against the glut of low-cost natural gas-fired electricity on the market.[74]

❸ Turn to the sun: promote renewable energy.

Efforts at fighting pollution, or any environmental problem, are ultimately constrained by the growing number of people. Emissions are rapidly rising in the developing world because of rapid economic and population growth there. Even so, industrial nations continue to put out nearly three-fourths of the world's atmospheric pollution and, for the next several decades, will remain the leading polluters of the atmosphere. Population growth here has much bigger impacts than population growth elsewhere: a typical North American pollutes ten times as much as an African.[75]

❹ Fight population growth. Read *Misplaced Blame: The Real Roots of Population Growth* (see inside back cover) for more ideas.

No matter how successful we are at slashing emissions, pollution already added to the atmosphere makes some degree of global warming inevitable. Minimizing our impact on the atmosphere must be our first order of business, but we also need to prepare for the inevitable. Floods will be less destructive where wetlands have not been paved over and logging roads don't crisscross mountains like spaghetti. Windstorms will topple fewer trees in forests that don't have sharp-edged clearcuts. Species will migrate more easily across landscapes that have not been fragmented by roads, clearcuts, or vacation homes. For these reasons, climate change argues for even greater protection of remaining natural areas and the diversity they harbor.
⑮ Protect and reconnect what's left. Old-growth forests, bogs, and grasslands also lock up huge stores of carbon in their plant life and soils. We want that carbon to stay there, not add to the pollution of the atmosphere.

FUNDAMENTAL SHIFTS in scores of policies and personal choices will be needed to cut deeply into greenhouse gas production. Yet most of the reforms would make sense even if the climate were not a concern. Phasing out subsidies to climate-damaging activities such as driving, old-growth logging, and aluminum smelting would cut tax bills as well as pollution. Reversing sprawl would make our communities more livable and save time, productivity, and lives now lost on our highways. Shifting the tax base away from income and sales and toward energy use or greenhouse gas emissions would spur employment while reducing waste.[76]

The economic benefits will be greatest for those who recognize that replacing the world's massive fossil fuel industry with the efficient use of renewable energy will be one of the largest growth industries of the coming century. Northwest companies—including more than 1,000 firms in the energy efficiency business—already have much of the expertise needed to help the world design a climate-friendly economy. Ballard Fuel Cells in Vancouver, B.C., is the world's leading designer of fuel cells, energy-efficient replacements for internal-combustion engines. Trace Engineering of Arlington, Washington, is the world's largest producer of electronic converters for solar and wind power systems. Siemens Solar in Vancouver, Washington, makes one-fifth of the world's photovoltaic crystals. Providing these technologies to a world hungry for them is necessary to fight climate change; it could also be a tremendous boost for the region's economy.[77]

Investments in energy efficiency simultaneously reduce the economy's fuel costs and its carbon dioxide emissions. Economists debate how much and how quickly energy efficiency can be improved cost-effectively. Some industry lobbyists argue that any improvements in efficiency will be expensive. But most economists agree that at least 10 to 30 percent emission reductions can be achieved at no cost or at a profit (and that's excluding the unfathomable costs that rapid climate change could impose on the world). Some studies suggest that energy consumption could be profitably reduced by half or more. Wherever the truth lies, it's worth noting that industry complaints about the cost of environmental proposals almost always prove to be greatly exaggerated. In the fight over the revised U.S. Clean

Air Act of 1990, coal-burning utilities' estimates of the cost of reducing sulfur emissions from their power plants exceeded the actual cost by a factor of *ten*.[78]

But what if coal and oil lobbyists are right, and efficiency investments end up costing more than they save? Does that mean we shouldn't act? Isn't it worth something to protect ourselves and our children? Isn't that why we gladly pay for health insurance, child seats, and fire departments? I know I would gladly pay to keep hemlocks from disappearing from the Oregon Cascades and salmon from disappearing from overheated streams. I'd pay to keep rivers from flooding my neighbors' homes and the Pacific Ocean from engulfing entire island nations. To be honest, it amazes me that anyone would argue we shouldn't do these things. Our pollutants have set a fire under our global home. When your home is on fire, you don't ask if you'll make or lose money on it. You put the fire out.

THE SCENE IS FAMILIAR to anyone who has read a magazine or watched TV in the 1990s. A sparkling sport utility vehicle tears across a remote and scenic landscape under a beautiful sky. Sometimes the four-by-four is speeding across a red-rock desert at dawn; sometimes it's barreling through a pristine mountain stream. Every ad markets the beauty of the outdoors to appeal to prospective buyers' fantasies of escaping traffic and urban hassles and perhaps of themselves as rugged adventurers. In reality, of course, these gas-guzzling machines are hastening the undoing of the outdoors and rapidly polluting those beautiful skies. That's not even mentioning what plowing through a stream will do to its water or its fish: according to Chrysler polls, 97 percent of Jeep Grand Cherokee owners never leave paved roads.[79]

The environmental undertone of these ads—like the Chevy ads that quote Thoreau or catalogs selling Eddie Bauer Ford Explorers that ask customers to donate a dollar to plant a tree—irks me. But then, why *shouldn't* gas guzzlers be named after places like Tacoma or the Yukon—places that will be hard hit by global warming? Maybe it's just a subtle form of truth in advertising. Why *shouldn't* automakers name their products—like the Cirrus, the Breeze, or the Aurora—after atmospheric phenomena? With all the carbon dioxide they put out, they *are* atmospheric phenomena.

One ad stands out above the rest. The gleaming four-by-four splashes through the rocky surf. Breakers curl, and a dark, forested shoreline rises in the background. The fate of our warming world was probably the furthest thing from the copywriter's mind, but sometimes you find wisdom where you least expect it. "Careful," the Infiniti ad reads, "you may run out of planet."

NOTES

The Northwest's greenhouse gas emissions, discussed throughout the book, were calculated by NEW. Fossil fuel–based CO_2 emissions are documented in endnote 4, other emissions in endnote 58.

1. J. Jouzel et al., "Extending the Vostok Ice-Core Record of Paleoclimate to the Penultimate Glacial Period," *Nature*, July 29, 1993; Intergovernmental Panel on Climate Change (IPCC), Working Group I, *Climate Change 1995: The Science of Climate Change* (Cambridge, U.K.: Cambridge Univ. Press, 1996).
2. Ten warmest years from Christopher Flavin and Odil Tunali, *Climate of Hope: New Strategies for Stabilizing the World's Atmosphere* (Washington, D.C.: Worldwatch Institute, 1996); downpours from Thomas R. Karl et al., "Trends in High-Frequency Climate Variability in the Twentieth Century," *Nature*, Sept. 21, 1995; sea level rise from IPCC, Working Group I, op. cit. note 1; earlier spring from R. B. Myneni et al., "Increased Plant Growth in the Northern High Latitudes from 1981 to 1991," *Nature*, April 17, 1997.
3. Amy Snover, "Impacts of Global Climate Change on the Pacific Northwest," presented at U.S. Global Change Research Program (USGCRP) and Office of Science and Technology Policy (OSTP), Pacific Northwest Regional Climate Change Workshop, Seattle, July 1997, also at tao.atmos.washington.edu/; IPCC, Working Group I, op. cit. note 1.
4. Regional greenhouse gas emissions were estimated based on NEW's calculations for B.C., Idaho, Oregon, and Washington, home to 93 percent of the Pacific Northwest's people. Fossil fuel–based CO_2 emissions were derived from energy consumption data in *State Energy Data Report 1994* (Washington, D.C.: U.S. Dept. of Energy [DoE], Energy Information Administration [EIA], 1997), also at www.eia.doe.gov; *Detailed Energy Supply and Demand in Canada* (Ottawa: Statistics Canada [StatsCan], 1960–1977); and *Quarterly Report on Energy Supply-Demand in Canada* (Ottawa: StatsCan, 1976–1995). Energy data multiplied by 99 percent carbon combustion rate and by carbon-per-unit-

of-energy factors from A. P. Jaques, *Canada's Greenhouse Gas Emissions: Estimates for 1990* (Ottawa: Environment Canada, 1992); and from EIA, *Emissions of Greenhouse Gases in the United States, 1987–1992* (Washington, D.C.: 1994). Statistics do not include fuel burned outside the region for electricity used inside the region. Before 1978, data are available only for the combined B.C., Yukon, and Northwest Territories region; we estimated that B.C. energy consumption was 89 percent of the total. Other greenhouse gas emissions cited in note 58. Population data from U.S. Dept. of Commerce (DoC), Census Bureau, *Current Population Reports*, series P-25, nos. 460, 957, and 1106 (Washington, D.C.: 1971, 1984, and 1993); DoC, Census Bureau, "Resident Population Estimates of the States, July 1, 1990 to 1994," www.census.gov/; *Revised Intercensal Population and Family Estimates, July 1971–91* (Ottawa: StatsCan, 1991); other Census Bureau and StatsCan sources; and *World Resources 1994–95* (New York: Oxford Univ. Press, 1994).
5. U.S. Congress, Office of Technology Assessment, *Preparing for an Uncertain Climate*, vol. 1 (Washington, D.C.: Government Printing Office [GPO], 1993), states that 60 percent of annual river flow in Washington is snowmelt.
6. IPCC, Working Group I, op. cit. note 1.
7. Nathan Mantua, "Observed Climate Variability in the Pacific Northwest," presented at USGCRP and OSTP, op. cit. note 3; Henry G. Hengeveld, "The Science of Climate Change," in Eric Taylor and Bill Taylor, eds., *Responding to Climate Change in British Columbia and Yukon* (Vancouver: Environment Canada and B.C. Ministry of Environment, Lands and Parks, 1997).
8. R. Berner, "The Rise of Plants and Their Effect on Weathering and Atmospheric CO_2," *Science*, April 25, 1997; IPCC, Working Group II, *Climate Change 1995: Impacts, Adaptations and Mitigation of Climate Change: Scientific-Technical Analyses* (Cambridge, U.K.: Cambridge Univ. Press, 1996).
9. IPCC, Working Group I, op. cit. note 1; "What's New about Global Warming in the Pacific Northwest?" U.S. Climate Action Network, Washington, D.C., 1997; Robert L. Peters and J. P. Myers, "Preserving Biodiversity in a Changing Climate," *Issues in Science and Technology*, winter 1991–92.
10. Snover, op. cit. note 3; Hengeveld, op. cit. note 7.

11. This chapter relies heavily on local climate scenarios developed by scientists at the University of Washington's Joint Institute for the Study of the Atmosphere and Ocean (JISAO) and at Environment Canada, based on global climate models. See Snover, op. cit. note 3, and Taylor and Taylor, eds., op. cit. note 7.
12. Robert G. Fleagle, "Policy Implications of Global Warming for the Northwest," *Northwest Environmental Journal*, fall–winter 1991.
13. Nathan Mantua, "Trends in the Present Climate and Climate Variability in the Pacific Northwest," in Amy Snover et al., *OSTP/USGCRP Regional Workshop on the Impacts of Global Climate Change on the Pacific Northwest* (Seattle: Univ. of Washington JISAO, July 1997); dam flows from Snover, op. cit. note 3.
14. Fraser River Sockeye Public Review Board, *Fraser River Sockeye 1994: Problems and Discrepancies* (Ottawa: Public Works and Government Services Canada, 1995); Scott G. Hinch et al., "Potential Effects of Climate Change on Marine Growth and Survival of Fraser River Sockeye Salmon," *Canadian Journal of Fisheries and Aquatic Sciences*, Dec. 1995.
15. Philip L. Jackson and A. Jon Kimerling, *Atlas of the Pacific Northwest* (Corvallis: Oregon State Univ. Press, 1993); Melinda M. Brugman et al., "Glacier Related Impacts of Doubling Atmospheric Carbon Dioxide Concentrations on British Columbia and Yukon," in Taylor and Taylor, eds., op. cit. note 7.
16. Jay R. Malcolm and Adam Markham, *Climate Change Threats to the National Parks and Protected Areas of the United States and Canada* (Washington, D.C.: World Wildlife Fund, 1997); prediction from Daniel B. Fagre, "Understanding Climate Change Effects on Glacier National Park's Natural Resources," in M. J. Mac et al., eds., *Status and Trends of the Nation's Biological Resources* (Washington, D.C.: U.S. Dept. of the Interior, forthcoming); Fagre quoted in "Gore Visits Glacier Park to Warn of Global Warming," Associated Press, Sept. 2, 1997.
17. Brugman et al., op. cit. note 15.
18. Hal Coulson, "The Impacts of Climate Change on River and Stream Flow in British Columbia and Southern Yukon," in Taylor and Taylor, eds., op. cit. note 7; Snover, op. cit. note 3.
19. *Climate Change: State of Knowledge* (Washington, D.C.: OSTP, 1997); U.S. Geological Survey (USGS), "Idaho Flood Summary," geohazards.cr.usgs.gov/html_files/nlic/idahosum.htm, Jan. 3, 1997.

Idaho road repair from Sarah Doherty et al., "Global Warming and the Pacific Northwest: A Profile of the Vulnerabilities of Our Region to Global Climate Change," Atmosphere Alliance, Olympia, July 1997; Glenn Bohn and Lindsay Kines, "Storm Bill Nears $200 Million," *Vancouver Sun,* Jan. 2, 1997.
20. Coulson, op. cit. note 18; Snover, op. cit. note 3; Romain Cooper, "Floods in the Forest," *Headwaters* (Headwaters Environmental Center, Ashland, Ore.), spring 1997.
21. Dave Birkland, "Fishermen Hook Marlin off Coast," *Seattle Times,* Sept. 4, 1997; Diedtra Henderson, "Northwest Finds Itself in Hot Water," *Seattle Times,* Aug. 12, 1997; Kevin Trenberth and Timothy Hoar, "The 1990–1995 El Niño–Southern Oscillation Event: Longest on Record," *Geophysical Research Letters,* Jan. 1, 1996.
22. Quotations from *Torrents of Change* (video) (Eugene: Forest Service Employees for Environmental Ethics, 1996).
23. IPCC, Working Group II, op. cit. note 8. Lee E. Harding and Emily McCullum, "Ecosystem Response to Climate Change in British Columbia and Yukon: Threats and Opportunities for Biodiversity," in Taylor and Taylor, eds., op. cit. note 7.
24. Jerry F. Franklin et al., "Effects of Global Climatic Change on Forests in Northwestern North America," *Northwest Environmental Journal,* fall–winter 1991; Richard J. Hebda, "Impact of Climate Change on Bioclimatic Zones of British Columbia and Yukon," in Taylor and Taylor, eds., op. cit. note 7.
25. Franklin et al., op. cit. note 24; J. Leverenz and D. Lev, "Effects of Carbon Dioxide–Induced Climate Change on the Natural Ranges of Six Major Commercial Tree Species in the Western United States," in W. Shands and J. Hoffman, eds., *The Greenhouse Effect, Climate Change, and U.S. Forests* (Washington, D.C.: Conservation Foundation, 1987).
26. Franklin et al., op. cit. note 24; Paradise from "What's New about Global Warming in the Pacific Northwest?" op. cit. note 8.
27. Harding and McCullum, op. cit. note 23; Glenn P. Juday et al., "Assessment of Potential Global Warming Effects on Land Ecosystems in Alaska," presented at BESIS Workshop on Impacts of Global Change in the Western Arctic Bering Sea Region, University of Alaska–Fairbanks, June 3–6, 1997, also at *www-cgc.admin. uaf.edu/besis/papers/Juday.html*; Franklin et al., op. cit. note 24.
28. Franklin et al., op cit. note 24.

29. Peter K. Schoonmaker et al., eds., *The Rain Forests of Home: Profile of a North American Bioregion* (Washington, D.C.: Island Press, 1997); Juday et al., op. cit. note 27.
30. Irrigated cropland from John C. Ryan, *Hazardous Handouts: Taxpayer Subsidies to Environmental Degradation* (Seattle: NEW, 1995).
31. Peter M. Vitousek, "Beyond Global Warming: Ecology and Global Change," *Ecology*, Oct. 1994; fire ecology from Robert Devine, "The Cheatgrass Problem," *Atlantic Monthly*, May 1993.
32. IPCC, Working Group I, op. cit. note 1; Paul D. Komar, "Ocean Processes and Hazards along the Oregon Coast," *Oregon Geology*, Jan. 1992.
33. Figure 1 based on Eric Taylor, "Impacts of Future Climate Change on the Lower Fraser Valley of British Columbia," in Taylor and Taylor, eds., op. cit. note 7; Douglas J. Canning, *Sea Level Rise in Washington State: State-of-the-Knowledge, Impacts, and Potential Policy Issues* (Olympia: Washington Dept. of Ecology, 1991); Dorothy Craig et al., *Preliminary Assessment of Sea Level Rise in Olympia, Washington: Technical and Policy Implications* (Olympia: Public Works Dept., 1993); Travis Hudson et al., "Regional Uplift in Southeastern Alaska," in Warren Coonrad, ed., *Alaska: Accomplishments during 1980* (Reston, Va.: USGS, 1982); and Komar, op. cit. note 32. East Coast comparisons from Robert J. Nicholls and Stephen P. Leatherman, "Adapting to Sea-Level Rise: Relative Sea-Level Trends to 2100 for the United States," *Coastal Management*, Oct.–Dec. 1996.
34. Highway 101 impact from Snover, op. cit. note 3; Richard A. Park et al., *Potential Effects of Sea Level Rise on Washington State Wetlands* (Olympia: Washington Dept. of Ecology, 1992); Leslie Beckmann et al., "Effects of Climate Change on Coastal Systems in British Columbia and Yukon," in Taylor and Taylor, eds., op. cit. note 7.
35. Richard E. Thomson and William R. Crawford, "Processes Affecting Sea Level Change along the Coasts of British Columbia and Yukon," in Taylor and Taylor, eds., op. cit. note 7; Beckmann et al., op. cit. note 34.
36. Thomson and Crawford, op. cit. note 35; Komar, op. cit. note 32.
37. Trenberth and Hoar, op. cit. note 21; Shannon Brownlee and Laura Tangley, "The Wrath of El Niño," *U.S. News & World Report*, Oct. 6, 1997.

38. IPCC, Working Group II, op. cit. note 8.
39. Quoted in Ross Gelbspan, *The Heat Is On: The High Stakes Battle over Earth's Threatened Climate* (New York: Addison-Wesley, 1997).
40. Carbon in gasoline from I. Boustead, *Eco-profiles of the European Plastics Industry, Report 10: Polymer Conversion* (Brussels: Association of Plastics Manufacturers in Europe, 1997). Manufacturing gasoline consumes 20.5 percent as much energy as gasoline generates when burned, according to Bill Driscoll, ICF Inc., Washington, D.C., private communication, Sept. 11, 1997. Emissions calculated using data from Mark Anderson, *Washington State Energy Use Profile 1960–1990* (Olympia: Washington State Energy Office [WSEO], 1992), and *State Workbook: Methodologies for Estimating Greenhouse Gas Emissions,* 2d ed. (Washington, D.C.: U.S. Environmental Protection Agency [EPA], Office of Policy, Planning, and Evaluation, 1995).
41. Howard Geller and Jennifer Thorne, "U.S. Carbon Emissions Climb 3.3% in 1996," American Council for an Energy-Efficient Economy, Washington, D.C., www.aceee.org/briefs/, July 31, 1997; A. Jaques et al., *Trends in Canada's Greenhouse Gas Emissions 1990–1995* (Ottawa: Environment Canada, 1997); per capita energy from *World Resources 1996–97* (New York: Oxford Univ. Press, 1996).
42. This chapter focuses on carbon dioxide; annual data on the other greenhouse gases are generally not available at the regional level. Although provincial and national greenhouse gas data are more current and comprehensive, the energy data used to calculate CO_2 emissions from the Northwest states are available only up to 1994. Jaques et al., op. cit. note 41; Figures 2 and 3 based on sources cited in note 4. The IPCC 60 percent reduction target is from J. T. Houghton et al., eds., *Climate Change: The IPCC Scientific Assessment* (Cambridge, U.K.: Cambridge Univ. Press, 1990).
43. Carbon dioxide in natural gas from Jaques et al., op. cit. note 41.
44. CO_2 per dollar based on personal income data from F. H. Leacy, ed., *Historical Statistics of Canada,* 2d ed. (Ottawa: StatsCan, 1983); *British Columbia Economic Accounts, 1985–1993* (Victoria: B.C. Stats, 1994); *Canadian Economic Observer* (Ottawa: StatsCan, 1993); *Historical Statistics of the United States, Colonial Times to 1970* (Washington, D.C.: Census Bureau, 1975); DoC, Bureau of Eco-

nomic Analysis (BEA), *Regional Economic Information System on CD-ROM* (Washington, D.C.: 1993); and BEA, *Survey of Current Business* (Washington, D.C.: Jan. and July 1994 and April 1995).
45. Cars are most fuel efficient between 30 and 50 miles per hour; outside that range, mileage drops noticeably, from Steve Nadis and James J. MacKenzie, *Car Trouble* (Boston: Beacon Press, 1993).
46. Alan Thein Durning, "Vehicles Outnumber Drivers in Northwest," *NEW Indicators* series, Jan. 11, 1995; "Greenhouse Gases in British Columbia," *Environmental Indicator Series* (Victoria: B.C. Ministry of Environment, Lands and Parks, 1996), also at *www.env.gov.bc.ca/sppl/soerpt/*.
47. Durning, op. cit. note 46.
48. Paul Rogers, "U.S. Fuel Efficiency Dropping," *San Jose Mercury News*, May 1, 1997; Matthew L. Wald, "U.S. Increasing Its Dependence on Oil Imports," *New York Times*, Aug. 11, 1997.
49. Tripled U.S. sales from Rogers, op. cit. note 48. Canadian light trucks from Jaques et al., op. cit. note 41, and from *1997 DesRosiers Automotive Yearbook* (Richmond Hill, Ont.: DesRosiers Automotive Consultants, 1997). U.S. vehicle registrations from *Statistical Abstract of the United States 1996* (Washington, D.C.: Census Bureau, 1996); sales expectations from Lesley Hazelton, "Ban the Sport Ute!" *Seattle Weekly*, Sept. 10, 1997.
50. Hazelton, op. cit. note 49. CO_2 emissions per mile (Figure 4) calculated by NEW based on Stacy C. Davis and David N. McFarlin, *Transportation Energy Data Book*, edition 16 (Washington, D.C.: DoE, 1996); carbon coefficients from *State Workbook*, op. cit. note 40, and from EIA, op. cit note 4; and Driscoll, op. cit. note 40.
51. Jaques et al., op. cit. note 41; Hazelton, op. cit. note 49; Keith Bradsher, "Collision Odds Turn Lopsided As Sales of Big Vehicles Boom," *New York Times*, March 19, 1997.
52. "Carbon Dioxide and Centralia," fact sheet, Northwest Environmental Advocates, Portland, no date.
53. BPA cuts from Northwest Conservation Act Coalition (NCAC), Seattle, *www.oz.net/ncac/*, Sept. 10, 1997.
54. Electricity consumption from *Statistical Abstract 1996*, op. cit. note 49. Home sizes from National Association of Home Builders, Washington, D.C., *www.nahb.com/sf.html*, Sept. 6, 1997; quotation from Julie V. Iovine, "Padding the Empty Nest: Couples

Build Bigger As Their Children Leave Home," *New York Times*, Sept. 4, 1997.
55. Rick Browning et al., "Impacts of Transportation on Household Energy Consumption, or Walk Your Talk," Browning-Shono Architects, Portland, April 1997.
56. Elliott A. Norse, *Ancient Forests of the Pacific Northwest* (Washington, D.C.: Island Press, 1990).
57. Mark Harmon et al., "Effects on Carbon Storage of Conversion of Old-Growth Forests to Young Forests," *Science*, Feb. 9, 1990.
58. Old-growth logging emission estimates based on methodology in Oregon Dept. of Energy (ODoE), *Report on Reducing Oregon's Greenhouse Gas Emissions* (Salem: 1995); conversion factor of 0.818 ton of timber per ton of old-growth tree bole from Harmon et al., op. cit. note 57; and ODoE's estimate (more conservative than Harmon et al.'s) that 25 percent of old-growth carbon is released and never recaptured after conversion to young forest. Old-growth logging estimates from B.C. Ministry of Forests (MoF), *Annual Report*, fiscal years 1993–94 and 1994–95 (Victoria: 1994 and 1995); David N. Larsen, *Washington Timber Harvest 1994* (Olympia: Washington Dept. of Natural Resources, 1996); Ken Denton, U.S. Forest Service, Seattle, private communication, Aug. 28, 1997; and Don Hicks, U.S. Bureau of Land Management, Portland, private communication, Sept. 23, 1997.

The long-term carbon impacts of logging in the drier inland Northwest are not well understood. These effects are likely to be smaller: trees in interior forests seldom reach the age or size of coastal old growth. NEW's estimate conservatively assumes that logging interior or "eastside" forests and previously logged forests causes no net emissions, based on William Ferrell, Dept. of Forest Science, Oregon State University, Corvallis, private communication, Aug. 11, 1995, and on Phil Comeau, MoF, Research Branch, Victoria, private communication, July 19, 1995. For a complete explanation of these calculations, contact NEW.

Figure 5 is also based on sources cited in notes 4 and 60. B.C. greenhouse gas emissions (except old-growth and fossil fuels) from A. Jaques et al., op. cit. note 41, and from A. Jaques, Environment Canada, Ottawa, private communication, July 7, 1997.

"Other CO_2" emission sources: Cement production from Hendrik van Oss, USGS, Washington, D.C., private communication, Sept. 9, 1997. Conversion of forests to sprawl, from Tracy Burrows, 1000 Friends of Washington, Seattle, private communication, Sept. 11, 1997; Charles Swindells, 1000 Friends of Oregon, Portland, private communication, Sept. 30, 1997; and Hal Swenson, U.S. Dept. of Agriculture, Natural Resources Conservation Service, Boise, private communication, Sept. 29, 1997.

Methane and nitrous oxide emissions based on *Washington State Agricultural Statistics 1995–1996 Annual* (Olympia: Washington Dept. of Agriculture, 1996); *1995 Oregon Agricultural Statistics* (Salem: Oregon Dept. of Agriculture, 1995); Idaho Dept. of Agriculture, "1994 Annual Report: Commercial Fertilizers," Boise, no date; *Agricultural Statistics, U.S. Dept. of Agriculture* (Washington, D.C.: GPO, 1996); James Kerstetter, Washington State Univ. Energy Program, Olympia, private communication, Sept. 17, 1997; *Report on Reducing Oregon's Greenhouse Gas Emissions* (Salem: ODoE, 1995); *Greenhouse Gas Emissions Inventory for Washington State, 1990* (Olympia: WSEO, 1994); and private communications with individuals at Idaho and Oregon Depts. of Environmental Quality and Washington State Depts. of Agriculture and Ecology.

59. Livestock numbers from *Statistical Abstract 1996*, op. cit. note 49; landfill regulations from James Kerstetter, Washington State Univ. Energy Program, Olympia, private communication, Sept. 5, 1997; Jaques et al., op. cit. note 41.

60. Elizabeth Cook, "Lifetime Commitments: Why Climate Policymakers Can't Afford to Overlook Fully Fluorinated Compounds," World Resources Institute, Washington, D.C., Feb. 1995. Smelter emissions calculated by NEW based on PFC emission rates from *State Workbook*, op. cit. note 40; CO_2 emission rates from Dean Abrahamson, "Aluminium and Global Warming," *Nature*, April 9, 1992. U.S. smelters assumed to operate at national average, 73 percent of capacity in 1994, based on Patricia A. Plunkert, "Aluminum," in *Minerals Yearbook*, vol. 1 (Washington, D.C.: U.S. Dept. of the Interior, Bureau of Mines, 1996).

61. Op. cit. note 60.

62. Erik Dolin, EPA, Washington, D.C., private communication, June 26, 1997; Michel Lalonde, *Canada's Climate Change: Voluntary Challenge and Registry Program* (Montreal: Alcan Smelters and Chemicals Ltd., 1996).

63. Table 1 based on sources cited in notes 4, 58, and 60; and on G. Marland and T. A. Boden, "Global, Regional, National CO_2 Emissions," in *Trends: A Compendium of Data on Global Change* (Oak Ridge, Tenn.: Carbon Dioxide Information Analysis Center, 1997).
64. John C. Ryan and Alan Thein Durning, *Stuff: The Secret Lives of Everyday Things* (Seattle: NEW, 1997).
65. Cook, op. cit. note 60.
66. Keith Kozloff and Roger Dower, *A New Power Base: Renewable Energy Policies for the Nineties and Beyond* (Washington, D.C.:World Resources Institute, 1993); Ryan, op. cit. note 30. Centralia based on Washington State Legislature, "Final Bill Report on Substitute House Bill 1257," *leginfo.leg.wa.gov/pub/billinfo/house/*, Oct. 14, 1997; Linda Brooks, Washington State Dept. of Revenue, Olympia, private communication, Sept. 16, 1997; and "Carbon Dioxide and Centralia," op. cit. note 52.
67. Michael Shelby et al., "The Climate Change Implications of Eliminating U.S. Energy (and Related) Subsidies," cited in "Lower Government Subsidies Can Help Economy While Reducing Emissions," *Global Change*, Sept. 1996.
68. Canada's action plan from Mauro C. Coligado, "Greenhouse Gas Emissions in British Columbia," in Taylor and Taylor, eds., op. cit. note 7. William K. Stevens, "U.S. and Japan Key to Outcome in Climate Talks," *New York Times*, Aug. 12, 1997; "Copps' Victory Is So Much Hot Air," *Vancouver Sun*, April 8, 1995.
69. *City of Portland Carbon Dioxide Reduction Strategy* (Portland: Portland Energy Office, 1993); Federation of Canadian Municipalities, "The FCM 20% Club," *www.fcm.ca_programs*, Sept. 25, 1997; International Council for Local Environmental Initiatives, Toronto, *www.iclei.org*, July 8, 1997.
70. John H. Cushman Jr., "Senate Urges U.S. to Pursue New Strategy on Emissions," *New York Times*, July 25, 1997; Carlotta Collette, "Oregon Is First," *Northwest Energy News* (Northwest Power Planning Council, Portland), summer 1997.
71. Seattle-sized savings from NCAC, op. cit. note 53.
72. "What's New about Global Warming in the Pacific Northwest?" op. cit. note 9; Environment Canada, "Canadian Passenger Transportation: 1996 Update," in *Canada's National Environmental Indicator Series*, *www1.sid.ncr.doe.ca/cgi-bin*, Oct. 7, 1997.

73. Kevin Bell, "The Unbearable Rightness of Green," *Cascadia Times,* Aug. 1997; wind from Kelly Redmond and George Taylor, "Climate of the Coastal Temperate Rain Forest," in Schoonmaker et al., eds., op. cit. note 29.
74. Jackson and Kimerling, op. cit. note 15.
75. Industrial nations' share of pollution from *Climate Change: State of Knowledge,* op. cit. note 19.
76. Ryan, op. cit. note 30; *America's Energy Choices: Investing in a Strong Economy and a Clean Environment* (Cambridge, Mass.: Union of Concerned Scientists, 1991).
77. Energy efficiency businesses based on Stan Price, Northwest Energy Efficiency Council, Seattle, private communication, Sept. 24, 1997; Siemens from Mike Nelson, Washington State Univ., Cooperative Extension Energy Division, Olympia, private communication, Sept. 24, 1997.
78. IPCC, Working Group III, *Climate Change 1995: Economic and Social Dimensions of Climate Change* (Cambridge, U.K.: Cambridge Univ. Press, 1996); "Global Climate Change: National Action," fact sheet, Natural Resources Defense Council, Washington, D.C., 1997.
79. Rogers, op. cit. note 48.

Here are some of the groups working locally to fight climate change.

Climate

Atmosphere Alliance
2103 Harrison, #2615
Olympia, WA 98502
(360) 352-1763
atmosphere@olywa.net

David Suzuki Foundation
Suite 219, 2211 W. 4th Ave.
Vancouver, BC V6K 4S2
(604) 732-4228
4228@vkool.com
www.davidsuzuki.org

Energy

Alternative Energy Resources
  Organization
25 S. Ewing, Room 214
Helena, MT 59601
(406) 443-7272
aero@desktop.org

Northwest Energy Coalition
219 1st Ave. S., Suite 100
Seattle, WA 98104
(206) 621-0094
ncac@nwenergy.org
www.nwenergy.org/ncac

Renewable Northwest Project
1130 SW Morrison St., Suite 330
Portland, OR 97205
(503) 223-4544
rnp@igc.apc.org

Transportation and Sprawl

ALT-TRANS
915 E. Pine
Seattle, WA 98122
(206) 325-9932
alt-trans@alt-trans.org
www.alt-trans.org/alt-trans

Alternative Transportation
  Centre
195 Terminal Ave.
Vancouver, BC V6A 4G3
(604) 669-2860
atc@wimsey.com
www.sustainability.com/best

Palouse-Clearwater
  Environmental Institute
P. O. Box 8596
Moscow, ID 83843
(208) 882-1444
pcei@pcei.org
www.moscow.com/pcei

1000 Friends of Oregon
534 SW 3rd Ave., #300
Portland, OR 97204
(503) 497-1000
info@friends.org
www.teleport.com/~friends

John C. Ryan is research director of Northwest Environment Watch, author of *State of the Northwest* and *Hazardous Handouts: Taxpayer Subsidies to Environmental Degradation*, and co-author of *Stuff: The Secret Lives of Everyday Things*. He has worked for local nonprofit groups in Indonesia and for Worldwatch Institute in Washington, D.C. John's home is in Seattle's Fremont neighborhood, but he practically lives on his bicycle.

Northwest Environment Watch (NEW) is an independent, not-for-profit research center based in Seattle. Its mission: to foster a sustainable economy and way of life throughout the Pacific Northwest—from southern Alaska to northern California and from the Pacific Ocean to the crest of the Rockies.

NEW's staff is Peg Cheng, communications director; Rhea Connors, office manager and volunteer coordinator; Alan Thein Durning, executive director; Kyle Halmrast, development and outreach director; John Ryan, research director; and Steve Sullivan, development and outreach assistant.